A CITY HERBAL

A CITY HERBAL

A Guide to the Lore, Legend,
and Usefulness of 34 Plants
That Grow Wild in the City

written and illustrated by

MAIDA SILVERMAN

 Alfred A. Knopf, New York, 1977

THIS IS A BORZOI BOOK
PUBLISHED BY ALFRED A. KNOPF, INC.

Copyright © 1977 by Maida Silverman
All rights reserved under International and Pan-American Copyright Conventions.
Published in the United States by Alfred A. Knopf, Inc., New York, and
simultaneously in Canada by Random House of Canada Limited, Toronto.
Distributed by Random House, Inc., New York.

Library of Congress Cataloging in Publication Data
Silverman, Maida. A city herbal.
Bibliography: p.
Includes index.
1. Plant lore. 2. Urban flora. 3. Botany,
Medical. 4. Wild plants, Edible. 5. Weeds. I. Title.
QK83.S58 1977   581.6'3   76-47930
ISBN 0-394-49852-6
ISBN 0-394-73297-9 pbk.

Grateful acknowledgment is made to Harper & Row, Publishers, Inc.,
for permission to reprint an adaptation of "Frijoles de Olla"
in The Cuisines of Mexico by Diana Kennedy.
Copyright © 1972 by Diana Kennedy.

The author also wishes to thank Farrar, Straus & Giroux, Inc. for permission to
reprint the quotation from Gigi by Colette. Copyright 1952 by Farrar, Straus
& Young, Inc. (now Farrar, Straus & Giroux, Inc.).

Manufactured in the United States of America

First Edition

For Abigail

CONTENTS

AUTHOR'S NOTE

There will probably be some concern on the part of the urban plant gatherer about the effects of air pollution on edible plants described in this book.

To the best of my knowledge, a major offender is lead from automobile exhaust. The subject is a complex one, because different plants as well as different parts of the same plant absorb lead and other pollutants at varying levels of concentration. As a general rule, plants to be used for food ought to be gathered as far away from obvious pollution sources as possible.

I must admit that I have been unable to pass up the bumper crop of excellently flavored Blackberries produced on bushes growing only a few feet from the edge of a heavily traveled highway near my home. Every summer we pick and eat several quarts, with no ill effects.

Soothing teas and facial washes recommended in this book are pleasant to use, but self-medication should not replace a physician's care for chronic or acute ailments.

ACKNOWLEDGMENTS

Many people helped with this book. Some shared special knowledge with me; others were supportive and encouraging. I am most deeply grateful to Grace Clarke, affectionately referred to as "our midwife," for it was she who set this book on its good path. Her enthusiasm and encouragement date back to the time when the book was hardly more than a gleam in my eye.

My husband, Martin, tasted and retasted many a recipe containing plants he never thought were edible, and ran the household when I was totally absorbed in writing, collecting, testing, and drawing. I am particularly thankful to Jane Garrett, my editor. Her patience, sensitivity, and special help during crucial times meant a great deal to me.

The courtesy, efficiency, and knowledge of Mr. Joseph T. Rankin, curator of the magnificent Arents and Spencer Collections of the New York Public Library, and of his assistants, Jeffrey H. Kaimowitz and Bernard F. McTigue, are gratefully acknowledged; it was a pleasure working with them. Mrs. Diane Schwartz, chief reference librarian of the splendid library of the New York Botanical Gardens, was very helpful, and enabled me to greatly expedite my research. I would also like to thank Kenneth J. Oberembt of the George Arents Research Library at Syracuse University. Through him I was able to locate a special work I had despaired of ever finding.

I am indebted to Larry Pardue, executive director of the New York Horticultural Society, and to British herbalist Judy Murray. They reviewed the manuscript for technical accuracy and made valuable suggestions, but are certainly not responsible for any errors or omissions. I thank copyeditor Mildred Owen, whose extensive knowledge of plants was a valuable plus.

I wish I could have thanked my beloved cousin Hannah Cohan for her years of love, encouragement, and faith in me. I thought of her often as I worked, and I believe this book would have delighted her.

I was very appreciative of and much impressed by the staff of Alfred A. Knopf, who were unfailingly cooperative and always available with excellent advice and assistance.

Last, but far from least, are the people I met gathering plants in vacant lots and roadsides of the city. Many had gathered and eaten these same wild plants in the "old country" and have been doing the same here; others remembered relatives doing so, and all generously shared their knowledge and recipes with me. To all these people, whose names I do not know, I am especially grateful, for their contributions were invaluable.

Kew Gardens, 1976

A CITY HERBAL

If a plant grows in abundance, it is a plant of virtue.

—Old Gypsy Proverb

You may not believe it, Gaston, but I often pick my best
camomile flowers in Paris, growing on waste ground,
insignificant little flowers you would hardly notice. But
they have a flavor that is unesteemable.

—Madame Alvarez, in Gigi, *by Colette*

INTRODUCTION

I have always lived in the city. Like many city dwellers, I have a fantasy garden in my head, planted with fruit trees, nut trees, flowers, berries, vegetables, and herbs. I once attempted a more modest version of this garden on my fire escape. Alas, it did not succeed, thanks to a family of hungry squirrels who ate the plants.

And so I turned my attention, perhaps out of desperation, to the wild green plants that grew in the city around me, and the more I studied and observed these plants, the more I came to love them. I found myself becoming involved in plant-lore, folklore, history, botany, pharmacology, psychology, witchcraft, magic, and superstition, nor does this exhaust the list; at one time or another in human history, almost all of the plants described in this book were highly esteemed, faithful companions who had served humanity since time out of mind. Some provided medicine and food; others were the source of beautiful colors for textiles. Many were used in magic and witchcraft, for in a hostile or at best indifferent universe, beset by events imperfectly understood, if indeed understood at all, it is hardly possible to over-estimate the important role plants once had in the attempts of men and women to manipulate or control the forces of the universe and the events of life.

To read about the plants I describe in this book is to find oneself becoming aware of the part they have played in humanity's timeless attempts to deal with the enigmas of birth and death, sickness and health, love and hate. Planting and harvesting, the fertility of the land, of women, men, and the useful animals, have been primary concerns of humankind; help was always needed to cope with the natural, unnatural, and supernatural mysteries of existence. And we are still just as preoccupied with these questions and answers. Practices that some of us dismiss as superstitious, far from being abandoned, are actually gaining ground, as a glance at the "occult" section of any library or bookstore will demonstrate.

The plants described in this book grow wild in cities. In vacant lots and waste places, along roads and highways, in the poorest of soils and from cracks in the side-walk where there seems to be no soil at all, they manage to flourish and thrive.

Few of our cities are beautiful, or even moderately attractive. Millions of us spend our lives in environments where the earth has been imprisoned and locked away forever under a sterile mantle of asphalt, concrete, and glass. Little space has been left for plants, which after all are not taxable and cannot be rented or otherwise be turned into money for someone. But we should be grateful for these plants. What hideous, barren places our cities would be without them!

To paraphrase Abraham Lincoln, God must have loved the common plants, since He made so many of them and generously endowed them with the blessings of hardiness, adaptability, and abundant fertility. Most of these plants have important characteristics that explain their success. They seem rarely if ever troubled by insects or disease. Some of them—Dandelion, Dock, and Chicory, for example—have long taproots. If the entire plant above the soil should be accidentally or deliberately cut off at the base, the root will just send up a new plant. Small bits of these roots if broken off and left in the ground will be able to produce plants.

The flowers of Mugwort, Lamb's-quarters, and Plantain, among others, are mentioned in the text as being "inconspicuous, minute, and densely clustered." They are replaced by equally inconspicuous, minute, and densely clustered seeds, and they have a germination rate that is truly astonishing: Lamb's-quarters can produce 100,000 or more seeds on a single plant. If conditions are favorable (and they usually are, apparently) these seeds sprout quickly, but they can remain dormant for years if necessary. A Wild Lettuce plant is capable of producing up to 50,000 seeds in a single season, and the seeds are able to remain dormant for forty years.

Candor obliges me to admit that some of the plants I love are "weeds" farmers spend billions of dollars annually to control. A "weed," according to my dictionary, is "any unsightly or troublesome useless plant, which grows in excessive, injurious or useless abundance, or any plant out of place." In that case, a crop of Dandelion plants deliberately being grown as a commercial crop for the sake of their leaves (and indeed, Dandelion greens *are* grown as a commercial crop) would not be "weeds," but a few stray rose bushes in that field growing among a commercial crop of Dandelions would be!

It makes me sad to hear these plants called "unsightly," for I do not believe there is such a thing as an unsightly or ugly plant. The prejudices that afflict *these* plants have caused them to be categorically dismissed, but I suspect that if the plants in this book were grouped—each in a plot by itself, in a garden designed after the manner of a Tudor herb garden, complete with brick paths—they would be very attractive. Few of us would even realize that they were the neglected and disliked "weeds" so common in the city. A minute Plantain flower, if viewed through an inexpensive hand lens, is as lovely and delicate as a snowflake. A Mullein leaf looks and feels like gray-green felt or wool—impossible to believe that it is really from a living plant—and the berries of Poison Ivy look like tiny, carved ivory beads.

An overabundance of some of these plants in pastures and cultivated lands causes serious economic loss, but many "weed" problems are a result of man's ruthless disruption of the balance of nature. As thousands of acres of our virgin native forests fell under the axes of the early settlers, a whole environment was destroyed and another created. Native woodland plants could not survive the destruction of their habitat and soon disappeared. Plants originally native to parts of the world where their habitat was similar to this newly cleared open land escaped cultivation soon after their arrival and spread rapidly. In 1672, John Josselyn wrote that Dandelion, Shepherd's-purse, Sow Thistle, Dock, Plantain, Mullein, and others had become well established and were flourishing in New England. Many aliens were brought over by the early colonists, for the sake of their usefulness, and were planted in gardens. But they were hardy, adaptable, and became naturalized quickly, with little if any competition. In some instances they were aggressive and crowded out the more desirable native species, but human agents are far more aggressive than any plant could hope to be and are responsible for the extermination of more native plants than any alien "weed."

Not too far from where I live are two large lakes. They and the extensive surrounding lands were the site of the 1964 World's Fair. I remember that every spring there was a particular boggy spot where we always found an exquisite little wild violet. It had a white flower and a narrow, lance-shaped leaf. One day it was decided, by some department in the city, to bring a pipeline through the area. In the process, the land was torn apart and turned into a wasteland of stones and mud.

That was several years ago. I used to revisit the place every spring, hoping that a few of these violets had somehow survived the general carnage, but I never did find any. The only plant growing there now is crabgrass.

There is a parking lot on the northwest corner of a street in New York City. The only place in the entire area that is not covered with asphalt or cement is a narrow strip of earth along the northern edge. In the spring, hundreds of green shoots emerge and soon become Dandelion, Clover, Melilot. As summer progresses, Mustard, Shepherd's-purse, Prickly Lettuce, Mugwort, and Lady's-thumb grow, and Bittersweet vines climb on the chain fence that separates the lot from the sidewalk. Toward fall, Goldenrod and Queen Anne's Lace bloom, and there are Milkweed plants, a few ailanthus trees, and several kinds of grasses.

I have chosen, from among the hundreds that grow all around us in the city, thirty-four plants that are special to me. They flourish and thrive in the cheerless places, and to me, all of them are worthy of notice, fascinating, and even beautiful, once you get to know them. Many are edible and delicious, and almost all are useful, in one way or another. My wish in writing and illustrating this book is to share these discoveries with you. If I succeed in doing so, it will make me very happy.

* * *

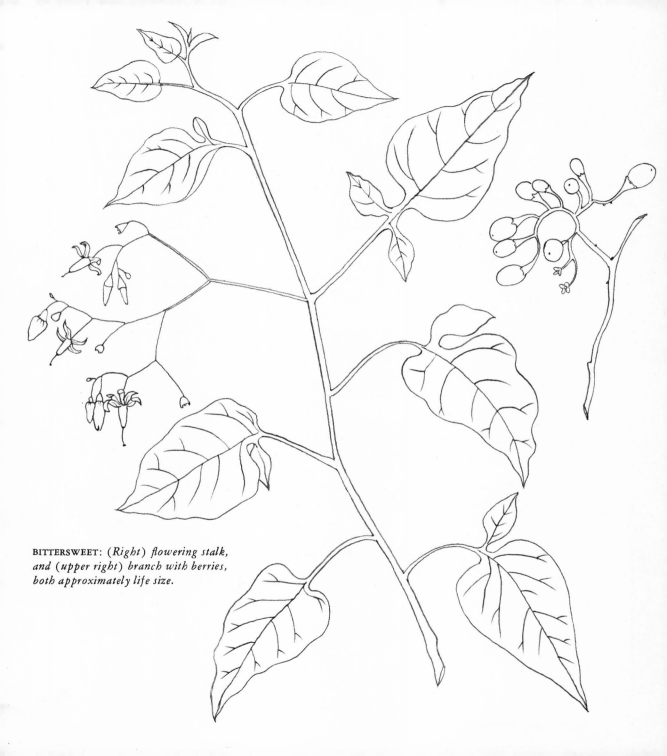

BITTERSWEET: (*Right*) *flowering stalk,
and* (*upper right*) *branch with berries,
both approximately life size.*

BITTERSWEET *(Solanum dulcamara)*

Folknames: Dulcamara, Felonwort, Nightshade, Shooting Star, Scarletberry, Violetbloom, Climbing Nightshade, Poisonberry, Dogwood.

Location: In cities Bittersweet can be encountered everywhere, usually growing among bushes or hedges, and along walls and fences.

Botanical Description: Bittersweet is a handsome, vinelike perennial plant. It may grow many feet in length, especially when supported by shrubs or fences. The leaves are very dark green, becoming a beautiful purple-bronze in the fall. On the upper stems, the leaves have from one to three winglike segments at their base; the lower leaves are heart-shaped. The purple flowers grow in clusters. Each flower has five reflexed petals and a bright yellow projecting stamen. The starlike appearance of the flowers does much to explain the folkname Shooting Star.

The berries start to appear early in August, small and green at first, growing somewhat larger as they become yellow, orange, and finally bright red. They resemble tiny cherry tomatoes.

Bittersweet is native to Eurasia and was introduced from Europe. It has become naturalized and grows throughout the United States, and is particularly abundant in the Northeast.

Historical Lore, Legends, and Uses: Bittersweet was called *Amaradulcis* (literally, "bittersweet") during medieval times. This name derived from the taste of the ripe berries, which, when eaten, first tasted sweet, then bitter. Some observers attributed this peculiarity to the root and twigs as well.

Exactly how dangerous or toxic Bittersweet actually is has been long debated. Some writers have claimed the ingestion of a few ripe berries to have been fatal; others insist that at worst, eating any part of the plant will cause mild to uncomfortable gastric distress. The poisonous principle solanine is found throughout the plant, but is believed most highly concentrated in the unripe fruit. Children should be taught to recognize this plant, so they will not accidentally eat the tempting-looking ripe berries.

Bittersweet is a member of the genus *Solanaceae,* and is closely related to the white potato, tomato, eggplant, sweet and hot vegetable peppers, petunia, tobacco, mandrake, thornapple, henbane, and belladonna, to name a few of the many important edible, ornamental, medicinal, and narcotic species of this genus. (These last three plants are frequently encountered as troublesome weeds in various parts of the United States.) The name *Solanaceae* is derived from the Latin "I ease," and many medicinal plants in the genus have long been referred to as "herbs of solace."

Several species closely related to Bittersweet are of legendary repute and their powerful narcotic and poisonous properties have been known and exploited since ancient times. Mandrake (*Mandragora officinarum*) is probably the most notorious. The earliest reference to it is found in the Old Testament. The plant is the subject of a poignant dialogue in the Book of Genesis (30:14) during which Rachel asks Leah for some of the plants as a fertility (or perhaps an aphrodisiac) charm: "Give me, I pray thee, of thy son's Mandrakes." Apparently, even as far back as biblical times, the plant had already become well established as an aphrodisiac, valuable in sexual matters; its popularity continued for centuries. The root was supposed to resemble the shape of the human body (there were even male and female plants!), and since the supply never equaled the demand, many a hand-carved substitute was sold to a gullible public.

During the Middle Ages it was believed that the plants shrieked horribly as they were dug up, causing madness or death to the gatherer, unless he resorted to any one of several fanciful and bizarre precautions. By 1636, however, John Gerard scorned the "many ridiculous tales brought up of this plant, wether of old wives or some runnagate Surgeons or Physike-mongers I know not." He detailed some very gamy folklore connected with mandrake, and ended by saying, "they are all and everie part of them false and most untrue." But some beliefs, especially where aphrodisiacs are concerned, die slowly if at all—the plant is still used in parts of southern Europe for the same purposes it was 2,000 or more years ago.

Belladonna, or deadly nightshade (*Atropa belladonna*), was so named because the noblewomen of Renaissance Italy put drops of juice pressed from the ripe berries into their eyes; this caused the pupils to dilate widely, making the eyes fashionably large and dark-looking. Belladonna happens to be extremely poisonous when taken internally, and it and the equally dangerous henbane (*Hyoscyamus niger*) were the all too effective ingredients in the murderous potions of the Borgias and other equally infamous poisoners.

Belladonna, henbane, and thornapple (*Datura stramonium*) were traditional ingredients in the notorious "Flying Ointment" of witchcraft legend. In preparation for their Sabbats, the witches rubbed this salve all over their bodies. In actual fact, visions could easily have been induced by inhaling the resulting narcotic and hallucinogenic fumes.

The sinister and deadly reputation of these plants was responsible for the suspicion with which edible members of the genus *Solanaceae* were greeted by Europeans.

The white potato, tomato, peppers, and eggplant are all from the New World, and were introduced into Europe by explorers. For years, these plants were considered unwholesome or actually poisonous.

Tomatoes were introduced into Europe during the early 1500s. Charles B. Heiser, Jr., writes that "the first tomato mentioned in Italy [c. 1544] was a golden color." At that time, the herbalist Matthiolus named it *pomi d'oro,* or "golden apple." *Poma amoris* and *pomme d'amour,* both meaning "love apple," are much later names for the tomato and there is uncertainty about whether these were corruptions of the name *pomi d'oro,* or derived from the belief that tomatoes were aphrodisiac.

By the seventeenth century, one writer claimed that the "love apple" *was* aphrodisiac, while another stated that it was called the "love apple" because its beauty was worthy to command love! The "love apple" was grown (by a very few) strictly as an ornament; the fruits were considered toxic.

In the United States, tomatoes were being used to make ketchup in 1779, and Thomas Jefferson grew them in his garden in Virginia. But they were not accepted as wholesome food in this country until the mid-nineteenth century, when they were finally deemed edible, and pronounced delicious.

The skin of the familiar white potato is poisonous when green (it then contains alkaloids similar to those found in Bittersweet), and the sprouts that grow from the "eyes" are poisonous also.

Today, thornapple (or jimsonweed, as it is sometimes called), belladonna, and henbane are the sources of several important drugs and are used extensively in modern medicine.

Bittersweet itself has had a long and controversial medical history. The great English herbalist John Parkinson observed in 1635 that the classical writers "did not remember this plant." Another writer mentions Bittersweet as having been used "since the time of Galen," the famous second-century Roman physician.

Bittersweet was reputed especially valuable in treating skin diseases, and for dissolving the congealed blood of bruises and other contusions. It was recommended as a cure-all for so wide a variety of illnesses that eventually its effectiveness was disputed.

William Coles, a well-known seventeenth-century herbalist, noted that Bittersweet was sometimes called Felonwort because "it took away felons [warts] which happen upon the joynts of the fingers." He reported that the leaves or berries of Bittersweet mixed with "rusty bacon" and applied to the felon were said by country people, "who are most subject thereunto, to be very successful for curing same."

Coles praised one remedy using Bittersweet as "a most excellent drink to open obstructions of the liver, gall, and spleen, and of good success in jaundice and cleansing women newly brought to bed." To accomplish all this, the leaves and tender branches of Bittersweet were added to three pints of white wine in a pot. The mixture was covered, set over a fire, and allowed to steep for twelve hours, then it was strained. The resulting liquid was drunk in quarter-pint doses morning and evening. This drink was also recommended for dropsy and difficult breathing. Coles declared that "it purgeth away watery humours very gently by urine."

Curiously, he nowhere mentions possible adverse side effects, nor does he attribute any poisonous properties to Bittersweet. John Parkinson, in his *Theatre of Plants* (a monumental herbal published in 1640, seventeen years before Coles's own herbal), specifically warned *against* taking Bittersweet internally in any form, noting that "it purgeth churlishly."

Most of the early writers were aware of the dangers of Bittersweet, and indeed, if one checks a few of the popular euphemisms associated with it, such as "removes menstrual obstructions," it may be that drinks containing Bittersweet were popularly used to induce abortion.

In our own country, one observer who had the highest esteem for Bittersweet was the nineteenth-century botanist C. S. Rafinesque. He traveled throughout the United States over a period of fifteen years, collecting and testing (often on himself) hundreds of medicinal plants, and was especially interested in the medicinal plants used by the native American peoples. He did not trust imported botanicals and particularly admired the herbs grown by the Shakers, who, he said, sold their products "cheap, fresh, portable and genuine." His great two-volume *Manual of Medical Botany of the United States* was published in 1826 and widely acclaimed.

Rafinesque, while aware of the caution with which Bittersweet should be administered, recommended the plant as valuable for purifying, obstruction removing (he is not specific about the type of obstruction), purging, and pain relieving. He called Bittersweet "a beneficial article in many diseases, now neglected by the chemical school, but adequate to produce all the good effects of sulphur, antimony and mercury in cases of gout, rheumatism, jaundice, asthma and all skin afflictions." Rafinesque believed that the American plants were stronger in action than the European, especially if grown in warm parts of the country.

Another nineteenth-century writer, a physician in fact, claimed to have personally observed twenty-one out of twenty-three cases of leprosy cured with Bittersweet administered internally and externally. Physicians of today would no doubt express surprise that the patients survived the treatment, and would even doubt the disease in question to be leprosy.

In China, crushed Bittersweet berries were mixed with vinegar and applied to sores to heal them, and in our own Ozark Mountains, until fairly recently, a popular, highly favored remedy for hemorrhoids was prepared by boiling Bittersweet berries with lard. In Westphalia (now a part of Germany) a decoction made from the entire plant was once a common drink and was supposed to prevent scurvy.

Bittersweet was long credited with magical and protective powers against witchcraft. An old German name for the plant was *Hynchkraut,* because it was thought to cure a disease of cattle called *die Hynch*—"swimming of the head." This affliction was attributed to witches, and shepherds hung wreaths of Bittersweet around the necks of their animals to remove the spell entirely or ward it off.

In England it was once believed that witches would come to the barns by night, steal horses out of their stalls, and ride them to their infernal meetings. In the morning, the owner would find the animal in its accustomed place, but exhausted and ill; this often continued until finally the poor beast died. To prevent the horse from being thus "hagridden," as it was called, Bittersweet and holly (another magical plant) were fashioned into garlands and hung around the creature's neck. This kept witches away and also cured the horse.

Herbal doctors still use Bittersweet occasionally to treat certain kinds of skin eruptions. They gather year-old twigs in late spring or early summer and dry them. The actual effectiveness of Bittersweet is doubtful, however, and the side effects possibly serious. The plant is considered poisonous and is no longer officially employed.

Tradition has long credited Bittersweet as a folk remedy for tumors and cancer, and a variety of Bittersweet gathered near Madison, Wisconsin, has been found to yield a tumor-inhibiting principle that has been named beta-solamarine and is being studied in laboratories at the present time.

Suggested Use: The flowering vines make an attractive indoor decoration in a vase by themselves or combined with other flowers.

BLACKBERRY *(Rubus species)*

Folknames: Bramble, Scaldhead, Fingerberry, Black Longberry.

Location: Road embankments, parks, along fences of vacant lots, and waste areas.

Botanical Description: Blackberry canes (as the stems are called) may be trailing, arched, or upright. They are quite flexible and sometimes bend down to the ground and root at the tip. Upright canes may grow more than six feet tall. The compound leaves are dark green with three- or five-lobed leaflets. These are variable in size and shape and have toothed edges. The illustration is a typical example, and both three- and five-leafed specimens grew on the same plant.

Blackberry blooms from mid- to late June. The masses of white or pale pink flower clusters make the plants conspicuous and easy to spot and remember for future berry picking. In the eastern part of the United States, Blackberries start ripening toward the middle of July. They are small, green, hard, and sour at first, becoming larger, and when fully ripe, juicy and sweet. Ripe and unripe berries frequently appear on the plants at the same time.

More than two hundred species of Blackberry are found in the United States. Some are low, trailing, and vinelike; others are large bushy plants. In the Northeast, purple-black berries are most common, but other varieties may be red or even white when ripe. Everyone loves the delicious fruit, and Blackberries of one kind or another can be found throughout the United States.

Blackberries are perennials that reproduce from seeds. The bushes are usually armed with thorns best described as vicious; they seem to deliberately reach out and grasp the clothing and skin and can inflict painful scratches on a careless picker. The thorns themselves have a nasty tendency to break off at the tip and become embedded in the skin. The eating more than makes up for the picking, however. The only other way to obtain Blackberries is to buy them, and they can be purchased only at gourmet greengrocers for a brief time during the summer, at a cost of about two dollars for a mere half-pint!

Historical Lore, Legends, and Uses: The word *bramble* is said to be derived from the Old English word *brymble,* meaning "prickly," and *bramble* can also mean "any thorny bush." Another source explains the word as coming from the Anglo-Saxon word *bræmel,* itself derived from an older word, *brom,* meaning "broom." In earlier times, the thorny branches of the Blackberry tied to a stick were used to make a broom for sweeping. In England, the word "bramble" is used as a verb; the expression "going brambling" means going Blackberry picking.

The Blackberry has long been appreciated for the taste of the ripe fruit and valued for its medicinal properties. Many writers did not even bother with a botanical description of the plant, saying instead that "it is so well known it needs no description," or "it grows in almost every hedge." It is difficult to overestimate the faith people once had in the healing powers of this plant. The astringent and binding properties were familiar to all who wrote about Blackberries and all parts of the plant—leaves, roots, flowers, and ripe and unripe berries were used. Preparations containing Blackberry were used to treat diarrhea, dysentery (often called "bloody flux"), various stomach disorders, and were believed valuable for healing irritations of the mouth and throat. Eating young shoots was even credited with fastening loose teeth in the gums! Infusions of the roots and leaves and syrups prepared from the berries added to wine were the usual methods of administering.

The Leechbook of Bald, a tenth-century Anglo-Saxon treatise on plants and herbal remedies, recommended preparations of Blackberry. For "flux in women" a tea was made from the berries and drunk for three days while fasting. For "heartache" the fresh leaves were pounded and laid over on the wound.

One Tudor herbalist recommended taking Blackberry juice mixed with wine and honey for "the passions of the heart." He observed, "The sweet ripe fruit is very effectual, besides the facility and pleasantness in taking." Another herbalist, Dr. William Coles, prescribed Blackberry as a remedy for "heartburn, as some call it, which is a gnawing in the stomach from choler." ("Choler" is an old word meaning anger.) This particular writer was apparently aware of the connection between emotions like anger and physical illness, especially stomach and digestive disorders. He noted that "the distilled water of [Blackberry] branches, leaves, flowers and fruit is very pleasant in both taste and smell and is excellent for feverish persons." At the end of his extensive treatise on the virtues of Blackberry, Dr. Coles decided to include the following homily: "The people of Norway use their bramble against scurvy and other melancholy diseases, so that we may admire the wonderful wisdom of God, who has ordained to grow in every climate remedies for those diseases whereunto it is subject." The doctor was no doubt rebuking his fellow Englishmen and women, who at that time were abandoning their native medicines in favor of foreign imported herbs, which he believed were greatly inferior.

The young roots and the root bark of the older plants were most favored for medicinal use. These contained the greatest amounts of valuable tannic, malic, and citric acids, and thus produced the strongest tonic and astringent effect.

The dried or green leaves were used to prepare gargles and heal sores and irritations of the mouth and genitals. One seventeenth-century writer states that "the powdered leaves strewn on running sores heals them." A decoction of the leaves was also valuable for treating stomach upsets and women's ailments. An infusion of the unripe berries was highly esteemed for curing vomiting and loose bowels. A wash for the hair (the leaves boiled in lye!) cured head sores and made the hair black.

Home remedies for the digestive ailments that frequently resulted from drinking unwholesome milk or water and eating tainted meats were kept on hand until well into the twentieth century, and this is still done in rural areas. Every kitchen had a supply of dried Blackberry leaves, roots, and berries on hand, as well as Blackberry jam, cordial, and syrup. The Pennsylvania Dutch used the leaves, roots, and fruit to ease indigestion, and preparations of the root were valued for treating diarrhea.

In China, several varieties of Blackberry were described and employed medicinally. The Chinese believed the fruit strengthens the "virile powers" and increases the "yin principle," in addition to giving vigor to the whole body. Preparations of the young shoots were used to improve the complexion and treat colds and fevers.

Blackberry was a familiar medicinal plant to native Americans. The Cherokee Indians chewed the root to ease coughs and used cold poultices to relieve hemorrhoids. Delaware Indians made a tea from the roots, which they used to cure dysentery, and the Oneida, Catawba, and other tribes were familiar with the root and used it for similar diseases.

At one time, Blackberry root was an official drug listed in the *United States Dispensatory*. A fluid extract for treating diarrhea was listed as recently as 1955.

From antiquity down to the present day, an astonishing amount of folklore and superstition have become associated with the Blackberry. Classical Greek writers, aware of the genuinely useful properties of the Blackberry, were convinced that the plant had the power to cure the bite of poisonous creatures. This belief persisted for centuries. In addition, Blackberry was credited with having a special affinity for women. The origin of this belief is lost in time, but perhaps the fertile habit of the plant or its many-seeded fruit was associated with the hoped-for fertility of women. In any event, Blackberry has for centuries been used to treat the disorders connected with menstruation, conception, and childbirth. Usually taken as a tea, the berries and leaves are still used by women in many parts of the United States and Europe.

The speed and efficiency with which the Blackberry bush can transform an area of open land into a virtually impenetrable forest of menacing thorns has made a tremendous impression, as testified to by the wealth of examples found in folklore: Sleeping Beauty protected by the bramble forest for one hundred years; Rapunzel, whose lover falls into the bramble patch and is blinded by the thorns; and our own Br'er Rabbit are a few examples of many. The bramble is also said to be the Burning

BLACKBERRY: (*Above*) *fruiting branch,*
approximately life size,
and (*left*) *five-leaflet leaf from lower stem,*
greatly reduced.

Bush from which God spoke to Moses, and from which the Crown of Thorns that tormented Jesus was fashioned.

The Anglo-Saxons believed Blackberry had the power to undo witchcraft. Christian faith and a pagan belief in the magical power of numbers (especially 3 and 9) are combined in the following charm "against any evil rune [witchcraft] and for one full of elvish tricks" (under enchantment): "Take Bramble Apple [Blackberries], pound and sift them, put them in a pouch, lay them under the altar and sing nine masses over them. Then put this dust into milk, drip Holy Water three times upon it and drink every three hours."

Another prescription, "a good salve against the temptations of the Fiend: and for a man full of elfin tricks," required ten different herbs, bramble included. These were pounded together, boiled in butter, and the resulting broth was strained through a cloth and set under a church altar. Nine Masses were sung over it and then "the man was smeared on parts of his body."

A peculiar contrast to the pleasure derived from eating the fresh, ripe Blackberries was the superstitious belief, in many areas in the British Isles and France, that Blackberries were unwholesome and even dangerous to eat. Thomas Green, a nineteenth-century British herbalist, mentions that "in some places Blackberries are called Scaldberries, from their supposed qualities of giving a disease called Scaldhead to children who eat them in immoderate quantities, as children will, when they can obtain them."

In many parts of England there was a widespread belief among country people that it was unlucky to gather or eat Blackberries after Old Michaelmas Day (October 11). Legend had it that Satan was thrown out of heaven on that day, and fell into the bramble bush, whose thorns caused him great pain. Ever since that time, on the yearly anniversary of his fall, he "spoils" the berries by spitting or breathing on them. Anyone foolish enough to eat any after that day will have bad luck, become ill, or perhaps even die. In one district, Blackberries were not eaten at all until quite recently, it being believed that "the trail of the Serpent" was upon them.

Blackberries were ruled by the planet Mars, and to dream of them foretold the future. It was generally a bad omen if you dreamed you walked through a bramble patch. If the thorns pricked you, secret enemies would do you harm through a trusted friend; if the thorns drew blood, you were to expect serious business reverses. If you passed through unharmed, however, you would triumph over your enemies.

In Cornwall, Blackberry leaves were employed to heal burns in a magical ritual of ancient origin. Water from a holy well was poured into a basin, and nine Blackberry leaves were floated on the surface. One at a time, each leaf was passed over and away from the afflicted area, the operator saying three times over each leaf:

There came three angels out of the East.
One brought fire and two brought frost.
Out fire and in frost,
In the name of the Father, Son, and Holy Ghost.

A variant of this charm was to lay the leaves on the skin; as they fell off, so would the burns heal.

Perhaps the most fascinating belief of all is the legendary efficacy attributed to the "bramble arch." One sixteenth-century writer noted that "every one that hath seen it is able to say that it [Blackberry] shoots forth many long, ribbed branches, which by reason of their length and weakness bend down to the ground again, there many times taking root." It was widely held that certain diseases could be cured by having the patient creep or be passed through this natural arch. The belief in this cure extended to animals as well as humans. In Normandy, until the latter part of the nineteenth century, cattle were regularly herded through these natural arches to cure them of lameness.

The bramble arch was considered even more powerful if the original plant was growing on one person's land and the tip of the cane had rooted on the property of another. It was very bad luck to destroy one of these arches or deny sick persons access to them. In Essex, a person (usually a child) suffering from whooping cough would be made to crawl under the arch seven times from east to west, saying each time: "In bramble, out cough, here I leave the whooping cough." A small cross fashioned from the thorny stems was worn on the person to ward off the disease, or aid in the cure if one was already ill.

Children were passed through the bramble arch to cure rickets and skin diseases as well. Along the Welsh border, after the child had crawled or been passed through, he was given a slice of bread and butter. The child consumed half the slice, and the other half was left under the arch. It was hoped that a bird or animal would eat this remaining piece and take the disease unto itself; the animal would then eventually die and the child would recover. Minor complaints such as rashes and boils could also be cured by crawling under the bramble arch three times from east to west.

All these rituals were meant to strengthen the cure. The healing powers of Christianity, the magic force of certain numbers such as 3, 7, and 9 as well as the life-giving force of the east-to-west "sunrise turn" would, it was hoped, enhance the recognized healing virtues of the Blackberry.

Suggested Uses: A modern writer suggests rubbing crushed fresh leaves on insect bites and scratches to relieve them, especially those received from the thorns of Blackberry bushes!

An infusion of the leaves is said to be useful for soothing sunburn, and other minor burns. I have used a syrup of the ripe berries to relieve upset stomachs and nausea, and to soothe sore throats and coughs.

In addition to the delights of eating freshly gathered Blackberries with cream and sugar, baked into pies, or made into jam, here are some additional ways to enjoy them.

BLACKBERRY SYRUP

Put ripe berries into a large pot and mash *very gently* with a potato masher. Add water to barely cover the berries, and cook over low heat until the steam rises and they start to give up their juice. Remove from heat and strain to remove seeds. Measure ½ cup of white sugar or mild-flavored honey (such as clover) for each cup of Blackberry juice, mix together the sweetening and juice, and reheat slowly *only until sweetening is dissolved*. Try not to let the juice or syrup boil, since this tends to destroy the appealing fresh-berry flavor. Keep syrup refrigerated in jars. (It may also be frozen, and keeps very well both ways.)

Blackberry syrup may be poured over cake, pancakes, or ice cream, and stirred into fresh fruit compotes. A few spoonfuls may be added to hot tea, and a pleasant drink may be prepared by adding 1 cup of boiling water to about an ounce of Blackberry syrup, along with a slice of lemon and sugar or honey to taste. This is especially welcome on a cold day, or before retiring.

BLACKBERRY CORDIAL

Prepare syrup as for Blackberry syrup (preceding recipe), or if you wish a sweeter cordial, allow 1 cup of sugar to each cup of juice. You may spice the syrup with nutmeg, cinnamon, allspice, or cloves, the combination and amount to your own taste. (Crush whole spices yourself. Do not use store-bought powders.) Boil syrup gently for about 10 minutes, then strain out the spices. You may use ½ cup of vodka (or brandy if you prefer) for each cup of syrup, or equal amounts of spirits and syrup. Store in tightly covered bottles.

BLACKBERRY VINEGAR

Barely cover ripe berries with malt or cider vinegar, gently crush to allow juices to flow, and let stand three days. Strain, allowing juice to drip overnight. Measure juice, and add ½ cup of white sugar or mild honey for each cup of liquid. Mix together and boil gently until sweetening is dissolved. Skim off any scum that forms. When cool, bottle and cap tightly. A teaspoonful mixed with water is an excellent thirst quencher, especially welcome during feverish colds and coughs.

Allow 1 heaping tablespoon of dried Blackberry leaves per cup of boiling water, cover, and steep 10 minutes. Strain and add honey or sugar to taste. You can combine equal amounts of dried mint and dried Blackberry leaves—an excellent combination.

Dried Blackberry-leaf tea makes an excellent astringent wash for oily skin. Prepare as for Blackberry-leaf tea, using ¼ cup of leaves. Let cool to tepid, strain, pressing out all liquid, and pat gently on your skin.

To dry Blackberry leaves: Pick leaves nearest the growing tips (from midspring to midsummer) in the morning when the dew has dried off. Lay on screens or hang up in small bunches *out of the sun,* but in a dry, airy place. When thoroughly dry, they will be crisp and brittle. Strip the whole leaves carefully from the stalks, and put away in jars. The leaves should retain their green color. To best preserve the active properties of the plant, do not crumble the leaves until you are ready to prepare the tea.

BLACKBERRY TEA

To prepare a tea with the dried green Blackberries, use 1 heaping tablespoon berries. Pour boiling water over them, and allow them to steep 15 minutes. Strain, and add sugar or honey to taste. I have found this tea useful to ease mild stomach upsets.

To dry berries: Generally speaking, it is the *green, unripe* rather than the ripe Blackberries that are dried because the valuable astringent principle is strongest in the unripe berries. Make sure clusters are free from moisture when picked, and dry as described above for the leaves. They will take longer than the leaves, and will shrivel considerably. Put away only when thoroughly dry, or they will mold.

Blackberries are delicious in apple pie. Add about 1 cup or more if you wish to pie filling.

To freeze berries: Wash gently and let dry thoroughly before freezing in plastic bags. It is not necessary to thaw before using.

Blackberry plants are a fine source of natural dye. A terra cotta color can be obtained from the leaves, and purple-brown from the ripe berries. See the Appendix, page 171, for instructions on preparing dye.

BOUNCING BET *(Saponaria officinalis)*

Folknames: Soapwort, Latherwort, Soaproot, Fuller's Herb, Crow Soap, Sweet Betty, Wild Sweet William, Old Maid's Pink, World's Wonder, Sheepweed, Lady's Washbowl, Bruisewort.

Location: Very common in the city, often growing out of cracks in cement along roads, fences, vacant lots, railroad rights-of-way.

Botanical Description: An appealing perennial, Bouncing Bet often grows in large clumps. Single stems may grow to a height of two feet or more. The leaves are oval, pointed at the tips, have smooth edges, and are an attractive matte-green color. The flowers, pale pink or occasionally rose-colored (there are single- and double-flowering types), have a delicate fragrance that seems more pronounced at dusk. Bouncing Bet usually blooms from July to September, and I have seen it flowering well into October. Native to Europe, Bouncing Bet was brought to America by the early settlers. It has long since escaped cultivation and spread throughout the eastern United States.

Bouncing Bet contains saponins, compounds that produce a soapy froth when shaken with water. These saponins, taken internally, severely irritate the digestive system, causing vomiting and diarrhea.

Historical Lore, Legends, and Uses: Most of the folknames for Bouncing Bet are derived from the plant's long-recognized ability to form a soaplike lather. Bouncing Bet has been long valued for its cleansing properties. Medicinally, it has a history of being employed to treat skin diseases, and was once (erroneously) considered a specific cure for venereal diseases.

The Anglo-Saxons were familiar with Bouncing Bet and called it Soapwort. (*Wort* is an old Anglo-Saxon word meaning "plant.") They prescribed it for liver ailments, for which it was drunk mixed with beer. Leprosy was treated with poultices made from the crushed plant mixed with wholemeal and vinegar. For swellings in general, a poultice was made by mixing the crushed plant with barley and wine. For earache, a tenth-century Anglo-Saxon leechbook gives the following instructions: "Work a salve thus: pound onion, leek and Soapwort, place in a glass vessel with vinegar and wring through a cloth. Drop the moisture into the ear."

A later writer comments: "The country people in diverse places do use to bruise the leaves of Soapwort and lay it on the fingers, hands and legs, when they are cut, to heal them up again."

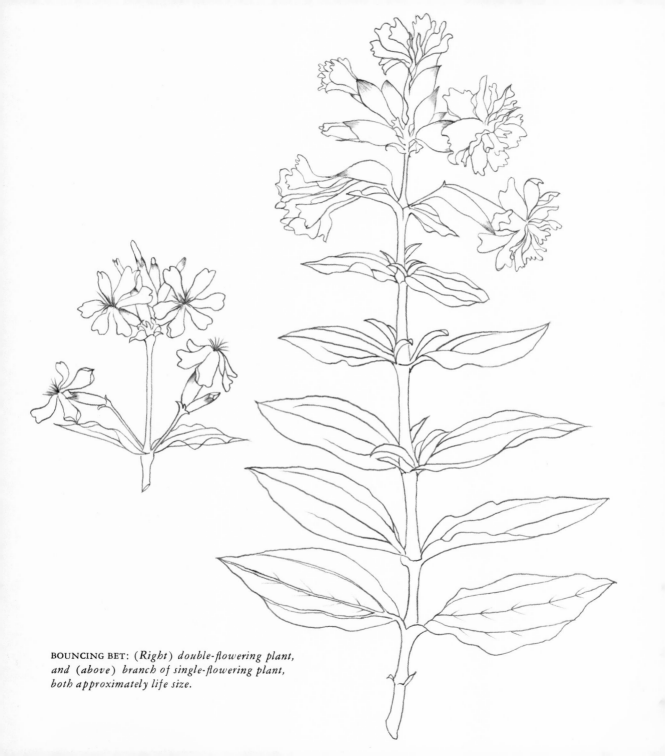

BOUNCING BET: (*Right*) *double-flowering plant,
and* (*above*) *branch of single-flowering plant,
both approximately life size.*

The Pennsylvania Dutch called Bouncing Bet *Seefa Schtuck*. They employed it to heal skin afflictions, as a soap substitute, and to produce a foamy head on beer! At the present time, some herbal doctors recommend a decoction of the crushed roots to relieve Poison Ivy rash, a treatment they claim is very effective.

To my mind, however, the most interesting aspect of Bouncing Bet is its long historical use as a cleaning agent, for which purpose it was employed long before the invention of soap as we now know it. For centuries, fullers (those textile workers whose job it was to clean, press, and generally prepare the newly woven cloth for the garment makers) grew Bouncing Bet in carefully tended plots and were dependent on it as a cleaning and soaping agent. In parts of the Near East, Bouncing Bet is still grown and used in this manner by the weavers of fine Oriental rugs.

Suggested Uses: Bouncing Bet has a beneficial action on the proteins of silk and wool. Pioneer women used it to clean their handmade lace; and preparations containing it are currently used by museums to restore color and sheen to fragile old textiles. The following recipe is recommended to anyone interested in the home cleaning of lace, brocade, stitchery, or other delicate old textiles.

BOUNCING BET "SOAP"

Gather fresh Bouncing Bet leaves, wash them, and crush them gently. Tie them up into cheesecloth bags in small batches. Put several bags at a time into a large pot full of soft water (rainwater if possible) and bring to a boil. When green foam starts to appear on the surface and the water turns greenish, remove from the heat and cool to tepid.

Meanwhile, dip the textile to be cleaned in cool soft water, swirling it gently until the water runs clear. This is to remove as much loose dirt as possible. *Do not squeeze or wring.* Then lay the wet textile on a board covered with clean soft fabric (such as an old sheet).

With a small, clean sponge dipped into the green foam, work the foam *gently* over the surface and into the textile, using a circular motion. Rinse the sponge as it becomes soiled and pick up fresh foam as needed. Wipe off foam and squeeze fresh tepid water through the textile, rinsing out all remaining foam and soil. Blot excess water and dry *flat* in an airy place, away from direct sunlight.

The delicately scented pink flowers of Bouncing Bet can be dried and added to potpourri mixtures.

To dry Bouncing Bet flowers: Pick freshly opened flowers and spread them, one layer deep, on trays. Dry away from the sun in an airy place. When thoroughly dry and crisp, store in jars away from the light, to preserve color and aroma.

BURDOCK *(Arctium lappa, A. minus)*

Folknames: Thorny Bur, Clotbur, Beggar's Buttons, Cockle Buttons, Sticktight, Lappa, Love Leaves, Bardana, Personata.

Location: Along roadways, vacant lots, and waste places generally.

Botanical Description: Burdock is a large, variable-looking plant. The first year's growth is a large rosette of ruffled, heart-shaped leaves. The second year, this rosette sends up a thick, many-branched stem from two to four feet high. The lower leaves are large, heart-shaped, and light green, with downy gray undersides. The upper leaves are egg-shaped and smaller.

Burdock blooms from mid-July through September. The flower petals are purple and bristly looking; they protrude from a cuplike receptacle thickly covered with hooked prickles. These receptacles and bracts, but not the petals, eventually become the familiar dry brown burs that, clinging to anything they brush against, distribute Burdock seed far and wide.

The tenacious taproot frequently grows straight down through the soil to a depth of three feet or more. It cannot be pulled out of the ground, but must be dug up. Any pieces that remain in the soil will become new plants.

Burdock is an alien. Originally from Europe, it was introduced by early settlers, quickly escaped cultivation, and is now found throughout the United States. Despite its name, Burdock is not a true Dock, but is a member of the *Compositae,* or Daisy, family. Some botanists recognize different species; to the lay person all are quite similar in appearance. Burdock is a biennial and reproduces by seeds.

Historical Lore, Legends, and Uses: Burdock was called "Burre Dock" in times past, and our common name, Burdock, may be a corruption of this. Other explanations have been suggested, however. The leaves of Burdock closely resemble those of broad-leafed Dock (*Rumex obtusifolius*), and the two plants often grow near each other. It may have been that Burdock was originally called "bur-dock," literally a kind of Dock with burs. Broad-leafed Dock was called "butter-dock" and "burre-dock" because farm women used to wrap their butter in the large cool leaves to keep it from melting on the way to market. Burdock is also mentioned as a plant used for this purpose, so Burdock may be a corruption of "burre [butter] dock."

John Parkinson, a great seventeenth-century herbalist, wrote extensively of Burdock. At one point he lapsed into what is best described as botanical theology, or, if you will, theological botany. He observed that classical Greek and Latin scholars believed Burdock had "degenerated" from a "good herb" into its present "offensively barbed" state. This, he declared, was "an error intolerable for Christians to aver, who may learn out of a truer school than Galen [a great classical physician] or the Heathens could, that the sins of Man caused God to curse the earth and caused it to produce [several kinds of prickly plants] and that each seed brings forth its own kind according to as God appointed at the beginning of Creation, or Man's Fall, at the least."

The genus name *Arctium* is believed to have descended from the Greek *arctos,* meaning "bear," alluding, perhaps, to the shaggy quality of the burs.

Shakespeare was familiar with Burdock and mentions it in *Troilus and Cressida.* Pandarus, in describing his relatives, says: "They are burrs, I can tell you, they'll stick where they are thrown."

During the seventeenth century, Burdock was commonly called *Personata,* and classical writers called it by the same name or a variant, *Prosopium.* This seemed to me a curious and intriguing name for a plant whose folknames mostly (and understandably) alluded to the sticky burs. *Prosopium* in Greek means "masked," and in Latin *personata* means "one who is masked." From this is derived the English word "persona," defined, according to Webster, as "a face mask used by actors in the sense of impersonating, hence a character or person in a play." Further research disclosed the fascinating information that the large heart-shaped leaves of the humble, now-ignored Burdock were once used as masks in ancient Greek drama to cover the faces of actors when they performed!

Burdock has a long medicinal history. The roots were used to treat dangerous fevers, skin diseases, and gout. Juice pressed from the root and drunk with old wine "wonderfully helps the bitings of serpents, and beaten with salt and laid on the place, quickly eases the pain." Burdock root "stamped and strained with Malmsey wine" strengthened the back, especially if there was added to it "the yolks of eggs and powdered acorns and nutmegs mixed together, and drunk first and last."

According to another herbalist, "the said roots are by long experience found to be very available against the plague and pestilential fevers by provoking sweat, and a decoction of the root in wine is singularly good for them that wheeze much, or are short-winded." The powdered, dried root was recommended for healing "naughty ulcers and fretting sores," and strong infusions of the fresh root were drunk to cure various kidney and bladder disorders. The fresh root could also be preserved with sugar for later use.

Burdock leaves used to be a major ingredient in an unguent called *Populneum.* Poultices of the leaves were popularly used to heal bruises and swellings, and country people boiled Burdock leaves in milk and used this dressing to soothe burns. Tenth-century Anglo-Saxon leechbooks mention Burdock; a poultice of the leaves was recommended to heal sore joints and old wounds.

During the Middle Ages a medicine to cure leprosy was prepared by pounding the leaves of Burdock with wine. This was heated as hot as could be tolerated, and applied directly to the lesions. After this treatment, afflicted areas were dressed with a salve made from honey, olive oil, and wine, and it was further recommended to the patient that "Burdock be your only drink." This medieval prescription ends on a confident note with the word "Proven."

It should be said that during this period in history, the fear of leprosy was so great that many harmless though unpleasant-looking skin diseases were probably mistaken for it. The above procedure was probably successful in healing some of these, and explains the adding of the word "Proven." This treatment, however, would not have cured true leprosy.

Various Delaware Indian tribes brewed a tea from the roots of Burdock. This was drunk to purify the blood, stimulate the system, and ease rheumatic pain, and was probably taught to them by the Europeans, since Burdock is not native. The Pennsylvania Dutch called Burdock *Gledda Wortzel* and brewed a tea from the year-old root. It was taken internally as a tonic, and used externally as a wash to get rid of dandruff and itchy scalp. They used Burdock seeds to treat kidney ailments and poultices made from the whole plant to help heal burns, superficial wounds, and minor skin irritations.

Burdock grows abundantly in northern and central China, where it has a number of folknames in various native dialects. The entire plant is highly esteemed for its medicinal virtues, similar, interestingly enough, to those ascribed to it in the West.

In addition to its medicinal properties, Burdock was considered a valuable potherb. In Italy, Scandinavia, and parts of France the young, tender leaves were gathered in the spring and added to salads; the roots were boiled and eaten with butter. In the United States, country people used to gather the tender leaf stems to eat. These were peeled and boiled, and usually eaten mixed with other greens.

Other uses have been suggested for Burdock. Thomas Green, a nineteenth-century writer, advised using mineral-rich Burdock as a fertilizer. "This herb," he reported, "burnt between the time it flowers and seeds, in a hole in the ground and not allowing the flame to escape; three pounds of ashes produce about one pound of a fine white alkaline salt as good as the best potash." The *Cyclopedia of Horticulture,* an esteemed twentieth-century reference work, recommends Burdock for use as an excellent foliage mass and screen.

Burdock came to be associated with a number of superstitions that have, through

BURDOCK: (*Above*) *leafy upper stalk, and* (*left*) *flowering stalk with burs, both slightly less than life size.*

the centuries, clung to it as tenaciously as its own burs stick to passers-by. The most fabulous, encountered time and again in the literature, is quoted verbatim from the writings of Dr. William Coles, a respected seventeenth-century physician: "The leaves of the Great Burdock are of great efficacy for bringing the matrix [womb] into its right place. If it has fallen down, the leaves laid to the crown of the head will draw it up, and if it should rise higher than is meet, as oftentimes it does, if laid on the soles of the feet, it fetches it down, and being applied to the navel, it suffers it to stir neither way." Burdock leaves were also tied to the stomachs of pregnant women to keep the child in position.

These procedures, so bizarre to us today, make sense if one knows something about the history of medicine. As far back as classical times and for many centuries thereafter, the anatomy of the human body and the etiology of disease were little understood. Physicians depended on theories (and there were many) to explain and treat the numerous afflictions the flesh was heir to. One of the most widely held of these was the "floating organ" theory. According to this philosophy, the organs of the body were not fixed in place, but were capable of "traveling" all over the body, thus causing a variety of illnesses.

The female uterus was believed especially prone to wandering about, and when this happened, all sorts of physical and mental disorders resulted. (The word "hysterical" is, in fact, derived from the Greek *hysteros,* meaning womb.) Obviously, then, physicians would be only too grateful for an herb like Burdock, that could be "depended on" as a remedy against "the displacement and suffocations of the matrix."

Burdock was a favorite herb of Venus, and therefore useful in love matters, to which this old charm attests: a girl picks a bur and names it for her lover. She then throws it against her skirt, or has a friend do so. If it sticks, he is true; if it does not, he is faithless. It was also averred that the young stalks of Burdock, peeled and eaten raw or boiled in the broth of fat meat, were not only pleasant to eat but "increased seed and provoked bodily lust." Contrariwise, however, was another claim that the root of Burdock boiled, soaked, and eaten with fat meat "cures venery" (lust).

According to an old New England superstition, Burdock leaves applied in the following manner would cure fever. They were bound to the wrists and ankles with the points down and the fever would "run out at the points."

Until fairly recent times, Burdock root was a recognized pharmaceutical in the United States. It is no longer officially listed, but in the 1917 edition of *Potter's Therapeutics* a fluid extract of Burdock (or dried Lappa Root, as it is called) is described as "excellent for promoting all kinds of secretions" with no irritating qualities. Burdock was a popular domestic remedy for rheumatism, gout, chest ailments, and persistent rashes. It was employed by physicians to heal hemorrhoids and chronic skin afflictions.

Burdock is listed as "Lappa Root" in the 1971 edition of the *British Herbal Pharmacopoeia.* It is used by herbal doctors in England and is considered to be a valuable remedy for certain skin diseases, especially when combined with Red Clover and Dock. It is employed as a diuretic and digestive stimulant, and preparations of the leaves, roots, and stems are used as a liver tonic. Recently, a French homeopath stated that a preparation of Burdock root drunk over a long period of time was capable of lowering the blood-sugar level in diabetics.

Children will enjoy gathering the green unripe burs, which, because they stick to each other, can be used to fashion baskets, mats, and amusing little figures of people and animals.

Suggested Uses

SAUTÉED BURDOCK STEMS

Gather a quantity of young Burdock stems in the early summer, before flower heads appear. Remove the leaves and peel off the thick, bitter rind. Dice the peeled stems, and plunge them into boiling water to cover. Boil 8 to 10 minutes, drain, and add fresh boiling water. Boil for 5 more minutes and drain well. The stems should be tender but crisp.

Melt some butter in a pan and add the diced Burdock. Sauté about 5 minutes and season with salt and freshly ground pepper.

Plain, diced, cooked Burdock can be served in a fine cream sauce or added to homemade vegetable soup. It can also be baked, with or without the cream sauce, in a buttered casserole. Sprinkle bread crumbs and freshly grated Parmesan cheese generously over the top, dot with butter, and bake at 375° until cheese is melted and bread crumbs are browned.

JAPANESE PORK WITH BURDOCK ROOT

Burdock root is called *gobo* in Japanese. It is a favorite vegetable, and has long been cultivated. It is available in Japanese food markets. It is dark brown and resembles a long carrot. Wild Burdock may be dug out of the ground for use in this dish. Serve it with boiled rice. Makes two generous servings. Will serve four if accompanied by other dishes.

3/4 pound pork chops with bones, 2 tablespoons Japanese soy sauce, 1 thick Burdock root, 1 tablespoon vinegar, 1-ounce piece fresh ginger root, salt to taste, 2 eggs, beaten, chopped scallions (optional)

Cut pork into 1-inch slices. Place bones in 2 cups water with soy sauce, and simmer about 30 or 40 minutes to make pork stock. Discard bones, set stock aside. While stock is cooking, prepare Burdock root. Wash it thoroughly, then peel outer rind with vegetable peeler. Score along length with a sharp knife, then peel off shavings with the peeler. Soak Burdock shavings in two changes of cold water, 20 minutes each time. Then place Burdock in a saucepan and pour boiling water over to cover. Add vinegar and simmer until tender, about 20 minutes. Drain, set aside. (Pork stock and Burdock root may be prepared in advance up to this point.)

Peel ginger root and slice into thin shreds. Bring stock to a boil, and add ginger and pork. Simmer until pork turns white. Add Burdock and bring quickly to a boil. Add salt to taste. When it is heated through, pour beaten eggs over all. When eggs are just set, remove from heat and sprinkle with chopped scallions if desired.

BUTTER-AND-EGGS *(Linaria vulgaris)*

Folknames: Toadflax, Fluellen, Flaxweed, Ramstead, Pattens-and-Clogs, Snapdragon, Churnstaff, Dragonbush, Brideweed, Toads, Yellow-rod, Lion's Mouth, Devil's Ribbon, Eggs-and-Collops, Devil's Head, Gallwort, Rabbits, Doggies, Eggs-and-Bacon, Buttered Haycocks, Monkey Flower, Calf's Snout, Wild Snapdragon.

Location: Common in the city and usually found in rather dry or sandy soil in lots, old yards, roadsides, and waste areas.

Botanical Description: Butter-and-Eggs forms great clumps of straight, thin-stemmed plants, occasionally branched, and from six to twelve inches tall. The stems bear many narrow blue-green leaves.

Butter-and-Eggs blooms from early July to late September. The distinctive flowers grow at the ends of the stems in compact spikes of yellow and orange. They resemble and are related to snapdragons (Butter-and-Eggs are smaller), having the same distinctive, lipped and long-spurred flowers. Originally from Eurasia, Butter-and-Eggs has become naturalized and now grows throughout the United States.

As an example of the way in which nature arranges for the pollination of a particular plant by a particular insect, there is hardly a better plant to study. The mouths of Butter-and-Eggs flowers remain closed and do not open unless forced to do so by bumblebees, honeybees, and other wild bees. The conspicuous orange spot on the lip attracts the attention of the bee and directs it toward two orange tracks inside the flower, which in turn lead toward the nectar contained in the spur. At the same time, pollen from the stamens (located on the "roof" of the flower) coats the bee's back. It has been observed that a bee can visit ten such flowers in about sixty seconds, cross-pollinating each time.

Historical Lore, Legends, and Uses: A plant with as many folknames as this must indeed have been well beloved. Several names refer to the animal faces the flowers bore a fanciful resemblance to: Lion's Mouth, Doggies, Rabbits, Calf's Snout, for example. On the other hand, the creamy yellow and orange colors of the flower reminded some of food: names like Butter-and-Eggs, Eggs-and-Collops, Buttered Haycocks, Eggs-and-Bacon are self-explanatory as well as imaginative. The bitter and disagreeable taste of the plant may account for a folkname such as Gallwort.

The plant is supposed to have been associated with toads, and some writers said this was because toads loved to sit under it, or that toads would eat it to heal themselves if they were ill. According to one source, however, the original name for Butter-and-Eggs was *Bubonium,* because the plant was once used to heal the buboes (inflamed lymph glands of the armpit and groin) associated with bubonic plague, a disease that ravaged Europe for centuries. This source claimed that a typographical error caused the word *bubo* to be printed *bufo,* which means toad. The error stuck, hence the names Toad and Toadflax. "Flax," the second part of that name, has its own interesting derivation. Before the flowers appear, Butter-and-Eggs greatly resembles the linen (flax) plant. This accounts for the generic name *Linaria,* derived from the Latin *linium.*

Butter-and-Eggs was called *Urinaria* by the old Latin scholars, who were familiar with diuretic properties of the plant. A later writer comments that "it is so well known unto all, that are never so little conversant with herbs, that it were almost needless to describe it."

Another observer believed Butter-and-Eggs to be the plant referred to in classical writings as *Herba Studioforum,* so called because scholars used to sweep their studies with a broom made from these plants. "Indeed," someone else said, "it seems to be a kind of Snapdragon, by its flowers."

The favored medicinal use of Butter-and-Eggs was as a diuretic. For this purpose, a decoction of leaves and flowers in wine was drunk. This drink was considered good in childbirth, helping to expel the afterbirth. The distilled water prepared from the entire plant was used to wash sores and ulcers to promote quick healing, and in addition was good for removing pimples and generally brighten the

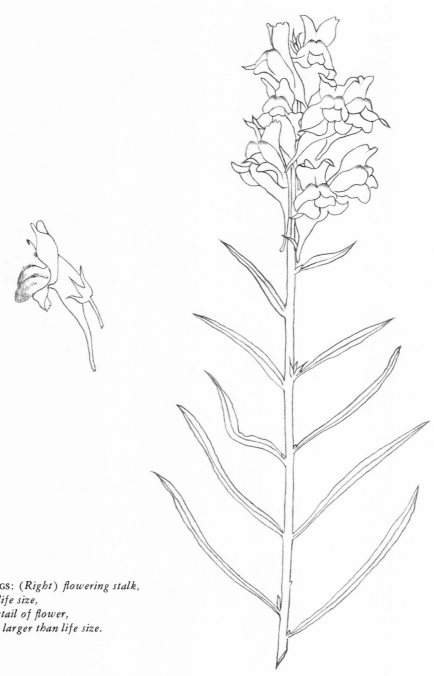

BUTTER-AND-EGGS: (*Right*) *flowering stalk,*
approximately life size,
and (*above*) *detail of flower,*
about one-third larger than life size.

complexion. An ointment for hemorrhoids was prepared by steeping the whole plant in oil. Roman women supposedly incorporated the juice of Butter-and-Eggs into a soap they used on their skin.

Even fowl benefited from this plant. One seventeenth-century herbalist says "The juice or distilled water put in the water chickens drink relieves them when they are drooping."

The Pennsylvania Dutch called this plant *Huns Blumma.* They prepared a decoction of the whole plant for use in kidney and bladder ailments. They used a poultice of the fresh plant to heal minor skin irritations and obtained a yellow dye from the flowers.

The close visual resemblance between the harmless Butter-and-Eggs and a poisonous plant is vividly related by the great herbalist John Gerard. In his *Herball* published in 1633, Gerard writes extensively about Butter-and-Eggs. He realized that it could easily be mistaken for the poisonous *Esula minor,* as the other plant was called: "The whole plant [Toadflax] before it comes to flower so much resembles *Esula minor* that one is hardly known from the other but by this old rhyme: *'Esula with milk doth flow / Toadflax without milk doth grow.'* " *Esula minor* was an old name for leafy or cypress spurge, a member of the genus *Euphorbiaceae,* and most of the plants in this genus have an acrid, milky juice varying from merely irritating to very toxic.

In times when Butter-and-Eggs was a popular internal medicine, improper identification could have proved very dangerous. I have often speculated, recalling Gerard's rhyme, if there did not at one time exist many more like it, and if there might even have been an oral tradition of rhymes, now forgotten and lost to us, that were used to teach plant identification in the days when books were uncommon and few could read.

Cypress spurge is, as a matter of fact, a fairly common weed in the city and is found throughout much of the United States as well. Before flowering, it is very difficult to distinguish from Butter-and-Eggs. Cypress spurge has small green flowers, however, and Butter-and-Eggs does not have a white sap. John Gerard's adventure with the poisonous *Esula* makes interesting reading: "Walking along the seacoast at Lee in Essex with a gentleman called Mr. Rich, I took but one drop [of *Esula*] into my mouth, which nevertheless did so inflame and swell my throat that I hardly escaped with my life. And in like case was the gentleman, which caused us both to take our horses and post for our lives unto the nearest farmhouse to drink some milk to quench the extremity of our heate, which then ceased."

Butter-and-Eggs was believed to possess supernatural powers and the ability to destroy evil spells. In ancient Scandinavia, the plant was said to kill flies. It was boiled in milk, and dishes containing this brew were set around the houses and outbuildings.

Suggested Uses

BUTTER-AND-EGGS SKIN WASH

A wash that will soothe and "brighten the skin" can be prepared as follows: pour 4 cups of boiling water over 1 cup of leaves and flowering tops of Butter-and-Eggs. Steep, covered, 10 minutes. Strain and cool to tepid. Pat gently on skin or add to the bath.

The dried flowers add pretty color to potpourri mixtures.

CHICORY *(Cichorium intybus)*

Folknames: Succory, Coffeeweed, Blue Sailors, Ragged Sailors, Blue Dandelion, Hendibeh.

Location: Along roadsides, in vacant lots and waste areas generally.

Botanical Description: Chicory is rather an angular and sparsely leafed plant, often reaching a height of four feet. Dark-green leaves form a rosette at the base. These lower leaves closely resemble those of the Dandelion (a close relative) but, unlike the Dandelion, are quite bristly or hairy. On the upper stems, however, the leaves are smooth, much smaller, and their bases clasp the stems.

Chicory flowers are numerous, usually in clusters of two or three and of a most beautiful blue. This color is rare in wildflowers and makes Chicory, a common plant in the city, all the more welcome. The plants bloom continuously in the Northeast from mid-June to September, with occasional flowers even later. It is a peculiarity of the flowers that they stay open only during a certain part of the day. So reliable is this regular opening and closing that Linnaeus, that renowned botanist and classifier of plants, grew Chicory in his famous "floral clock" in Uppsala, Sweden. In that latitude, the flowers could be depended on to open at 5:00 A.M. and close at 10:00 P.M.

Chicory is an alien. Originally Eurasian, it was introduced from Europe by the early colonists. It escaped cultivation, became naturalized, and now grows throughout the United States. Chicory is a perennial and reproduces from seeds. It develops a very long taproot that is most difficult to remove from the ground. New plants will readily develop from any pieces left broken off in the earth.

Historical Lore, Legends, and Uses: Chicory is a plant that has been known and used since antiquity. The Latin name *Cichorium,* or a variant of it, has become the name for the plant in almost every European language. This name has been traced back from Latin (the Romans enjoyed Chicory in salads), Arabic, and classical Greek to the ancient Egyptians. They called it *Chickoryeh* and were very fond of eating it. The biblical description of the feast of the paschal lamb (Exodus 12:8) reads, ". . . with bitter herbs they shall eat it." Chicory has been identified as one of these "bitter herbs," and to this day it is traditionally included on the Seder plate as part of the Passover meal, which is the oldest continually celebrated religious rite extant.

Chicory has been an esteemed medical plant ever since the Roman physician Galen called it "the friend of the liver" some 1,800 years ago. A syrup of Chicory, rhubarb, and oats was given to patients with liver ailments. In the seventeenth century, a physician-herbalist declared enthusiastically, "Chicory any way used as long as it is green, cools the heat of the liver and by special property strengthens it and opens obstructions thereof, for which virtues it deserves to be much esteemed. It is," he correctly observed, "a great preserver of health to have the liver temperate and unstopped. If this part do not rightly perform its office, things will not go well."

In addition to its specific use in liver diseases, Chicory was considered valuable for treating a variety of other ailments. A syrup of the whole plant was prepared with sugar and taken to cure insomnia. The bruised fresh leaves were applied externally for healing eye inflammations, and boiled in broth were good for strengthening the digestion of persons with weak stomachs. An infusion of the leaves was also used to reduce fever in children.

According to Dr. William Coles, Chicory root boiled in wine or water and drunk while fasting, "drives forth choleric and phlegmatic humours, helps dropsy and also those with an evil disposition in their bodies by reason of long sickness or evil diet." Herbalists recommended Chicory, or *Succory* as it was called, for "fevers, faintings, swoonings and *the passions of the heart."* A distilled water of Chicory or the juice pressed from it was good for pregnant women, and especially to soothe nurses' breasts "painful from too much milk."

There is more to this last suggestion than is immediately apparent: during the seventeenth century, a widely accepted medical philosophy known as the doctrine of signatures was in vogue. According to the theory, God, in His wisdom, had placed remedies close at hand to cure the afflictions of humankind. To aid in determining which herb would cure what disease, each plant was "signed" in a way that made its application easily determinable. An example of such a plant was Chicory; its milky white juice was interpreted as a "sign" that Chicory was a remedy for treating the difficulties of lactating women.

This curious philosophy, elements of which date back to classical times, contained many built-in ambiguities and obvious contradictions but was nevertheless a recognized basis for treating disease for many years. In fact, the doctrine of signatures, along with other mystical and occult ideologies, is presently undergoing a revival and has numerous adherents today.

Chicory was much appreciated as a cosmetic. Several old writers mention that a preparation of the gummy juice was used to smooth the "flanging haires" of the eyebrows, of all things! Fresh Chicory leaves crushed with vinegar and applied to the skin were a much-appreciated remedy for pimples and other blemishes. The leaves were also chopped, made into little round cakes, dried, then dissolved in rosewater when needed. Used as a skin wash, this mixture made the complexion "most fair" and also healed chapped skin. A decoction of Chicory with water and honey was applied to the breasts to (hopefully) make them round and firm.

Chicory root has long been popular in Europe for giving coffee a very dark color and a characteristic taste. Its presence is supposed to add flavor and correct the acidity and excessive stimulation of coffee drunk by itself. For this purpose the root is cleaned, dried, roasted until brown, then ground to a fine powder. Chicory lacks the caffeine and volatile oils peculiar to real coffee, but during World War II it was much used as a substitute when real coffee was unavailable.

In the latter part of the eighteenth and early nineteenth centuries, Chicory was even suggested as an American farm crop. At that time, various ethnic groups were substantial consumers of Chicory, for use with coffee. The imported root was preferred, however, and was a duty-free import at that time. In 1898 a protective tariff was placed on imported Chicory. Factories for processing it were built in this country, and farmers were encouraged to grow it as a crop. Unfortunately, the operation proved unprofitable and was eventually abandoned.

Coffee with Chicory was and still is popular in parts of the South, especially Louisiana, where it is called Creole coffee. A plantsman writing in 1946 observed: "Several thousand tons [of Chicory] are imported yearly, which could be found in one American county, such as Orange County in New York." Farmers would be only too pleased to have people dig the plants out of their fields and pastures. Cows love to eat it, but Chicory makes milk bitter and products derived from it unfit to eat.

CHICORY: (*Left*) *flowering stalk,*
and (below) basal stalk
showing portion of root,
both approximately life size.

Chicory happens to be a plant strikingly rich in lore and legend. It was considered an herb of Venus, and consequently was prized for love charms and potions. One kind of charm was fashioned from the root: a girl dug Chicory from the ground on one of the Apostle's Days (June 29 and July 25) with a golden knife. Worn as an amulet, it assured her of her lover's faithfulness. Chicory seeds were added (secretly of course) to the food or wine of a person whose love was desired.

Chicory is believed to be the famous "Luck Flower" of German mythology. Blue and starlike, the Luck Flower was born of lightning. Whoever possessed it could cause the very mountains to open, thus gaining entry to the subterranean caverns of the netherworld, where great treasures of gold and precious gems were to be found.

The possessor was warned that he must bring the Luck Flower back with him if he wishes to return safely to the outer world, but usually he is overwhelmed by his good fortune (or too greedy to bother) and he forgets the injunction. Suddenly, the mountains close in his path, making him a prisoner forever, or the rocks tumble down upon him and he is killed.

An old German folkname for Chicory was *Wegewarte,* meaning "watcher of the road." According to the legend, a maiden waiting for her lover to return from a long voyage went out every morning to watch on the road where she had bade him good-bye. Months passed, but he did not return. Each morning she would go out again, and each evening she would return, broken-hearted. Finally, exhausted with grief, she sank down by the wayside and died. A plant sprang up on the spot; opening in the morning and closing at dusk, the beautiful blue flowers of the *Wegewarte* continued the vigil of the maiden.

There is an old Silesian folktale that describes how Chicory (*Czekanka*) got its name: centuries ago, a sorceror had a beautiful blue-eyed daughter named Czekanka, who was betrothed to a handsome youth. He was murdered by a rival, and in despair, Czekanka killed herself at his grave. Her sorrow-stricken father changed her into a Chicory plant, with blue flowers in remembrance of her lovely eyes, so that she might forever remain beside the tomb of her beloved. The murderer then vengefully commanded a swarm of ants to destroy the plant, but the ants turned on him instead, attacking him fiercely and pursuing him until he fell over the edge of a cliff and was killed. From that time on, Chicory was called *Czekanka*.

Another legend explains why the Chicory flowers close at noon: long ago, the Sun fell in love with a maiden for the sake of her eyes, of an exquisite, starry blue. He begged the girl to marry him, but she thought herself too good for the Sun and rudely refused him. He changed her into the Chicory plant and commanded her to gaze at him forever, from sunrise to sunset, with flowers as blue as her eyes had been.

Despairingly, she called upon her mother, a powerful witch, to restore her to human form. This her mother could not accomplish, but she weakened the enchantment so that the flowers could have respite and close at noon, which they do to this very day.

Perhaps the oldest, most widespread, and surely the most fantastic legend associated with Chicory was the belief that it had the magical ability to open doors and boxes. The strongest locks and stoutest bars would immediately fly open at the first application of this powerful charm flower, which had the additional and convenient advantage of rendering its possessor invisible. To be efficacious, however, the plant had to be gathered on St. James's Day (July 25) at the stroke of midnight. It had to be cut with a golden knife in strictest silence, and if the person harvesting it uttered so much as a sigh, he would die immediately, or soon thereafter.

Today, Chicory is no longer gathered with a golden knife to open locks. Instead, modern herbal doctors dig it up with a garden trowel and use it to prepare a tonic for various liver disorders. They believe Chicory increases the secretion of bile. It has been demonstrated, however, that excessive or prolonged use of Chicory root can upset the digestion, so caution is advised for its use as a beverage.

Suggested Uses: Chicory grows abundantly in urban areas and deserves to be better known and used. Early June is a good time to start gathering the tender upper leaves; they are rich in vitamins and make excellent edible greens. Young leaves can be gathered in the spring and added to salads, cooked with other potherbs and eaten with butter and vinegar, or chopped and added to broth. Several varieties of Chicory have been developed commercially and are available in markets.

I have stopped to talk with (and join) many people, French and Italian especially, who throughout the spring regularly visit the empty lots and fields in the city to gather the young leaves of this abundantly available plant.

The following recipe (which reads quite poetically) is taken from a volume entitled *Works of Singular Ladies and Gentlemen* written in 1682 by Giles Rose:

Take the leaves of Wild Green Succory, cut them small, and put them into faire water and so let them lye two hours. Change your water 3 or 4 times, then swing it out very well, and Dish it up on a Plate and garnish it with any Thing, either red or white.

Here is a modern adaptation:

SPINACH, CHICORY, AND MUSHROOM SALAD WITH BACON DRESSING
*2 cups spinach and Chicory leaves per person, 2 strips bacon per person,
fresh mushrooms as desired, salad oil, wine vinegar, salt and freshly ground pepper*

Wash and thoroughly drain spinach and Chicory. If spinach leaves are very large, tear them into smaller pieces. Fry bacon until crisp, and reserve about 3 tablespoons of the bacon fat to add to the dressing. Slice mushrooms. Make a dressing with oil, some of the reserved bacon fat, vinegar, salt, and pepper. Toss with greens and mushrooms, and crumble bacon over. Serve immediately. This salad can also be made entirely with Chicory.

CHICORY ITALIAN STYLE
freshly gathered wild Chicory leaves, olive oil, garlic, salt and freshly ground pepper

Wash Chicory. Plunge into boiling water, cook 5 minutes, and drain. Taste a leaf. If it is too bitter, repeat procedure. Drain well. Heat olive oil, and finely chop 1 or 2 garlic cloves. Add to the oil but do not let garlic get brown. Add Chicory, and cook only long enough to heat through. Add salt and pepper to taste.

COFFEE WITH CHICORY

Roasted ground Chicory root is available in health food stores and gourmet food shops, where it is rather expensive. Coffee with Chicory is an agreeable Old World beverage, and if you wish, you can prepare the roots yourself. Dig the roots out (they are impossible to uproot by merely pulling) in late summer. Scrub them well, but do not peel. When they are quite dry, roast them in a slow (about 300°) oven until they are brittle and dark brown. The length of time will depend on the thickness of the roots. Break up into small pieces and grind as you would coffee beans. (A small electric coffee grinder is excellent for this.) Store in jars with tight-fitting lids. You can experiment by adding 1 part Chicory to about 4 parts coffee, or to your own taste. Serve Chicory coffee with hot milk and sugar.

CLOVER, RED *(Trifolium pratense)*

Folknames: Trefoil, Honeysuckle, Beebread, Clover Rose, Ladies' Posy.

Location: Red Clover grows everywhere in the city, especially in vacant lots, parks, and along roadways.

Botanical Description: Red Clover is a low-growing plant, generally between five and eight inches high. The compound leaves are divided into three leaflets, oval in shape and usually banded in the centers with a distinctive white chevron. The globe-shaped flower heads, like those of the closely related White Clover, are actually made up of tiny individual tubelike florets. The flowers are very attractive and a joy in the city, for they are large (sometimes more than an inch in diameter), fragrant, and range in color from a delicate pale pink to deep magenta purple. The blossoming season is a long one in the East, from May well into October.

Red Clover is an alien. A valued soil improver in the Old World, it was introduced from Europe and has become naturalized throughout our country. It is a perennial and reproduces from seeds.

Historical Lore, Legends, and Uses: Red Clover is usually thought of as a bee plant, and bumblebees particularly are nearly always seen flying around the flowers. Honeybees, on the other hand, will not work it as long as they will White Clover, and it is not an especially good source of nectar for them. Red Clover *does* happen to be a fine source of "green manure," however. It is often grown as a cover crop and plowed under. The nitrogen-fixing bacteria that grow in association with its root system greatly improve the soil.

Red Clover has been used for food and medicine. A decoction of the entire plant, seeds and root included, was (and still is) drunk by women to check excessive menstrual flow, and a tea of the flowers—dried are stronger than fresh—was drunk with honey to relieve cramps. The juice of Red Clover was even valued as a specific for eye diseases. In the words of one seventeenth-century physician, "It is a familiar medicine with divers persons to take away the *pin and web,* as they call it, from the eyes, by Signature." ("By Signature" meant that the oval-shaped leaflets, with their white spot in the center, resembled the human eye, and that Red Clover could thereby be expected to cure eye diseases.) The leaves and blossoms were boiled with lard and used as an ointment for cuts, bites, and venomous stings.

Country people in many places believed the Red Clover to have particular power to cure the bites of poisonous snakes and insects. They boiled the whole plant in water, washed the bitten place with it, then laid some of the freshly crushed leaves on the area.

In California, Red Clover was considered a food by many native Americans. They generally ate the fresh leaves before the flowers appeared. The Mohegans of Connecticut steeped the dried leaves in hot water to make a soothing tea for sore throats and colds, and other native American tribes used the plant to heal skin eruptions.

The Pennsylvania Dutch called Red Clover *Rhoda gae Blumma.* They brewed the dried blossoms into a tea as a remedy for croup and whooping cough. In the Ozarks, a strong tea of Red Clover blossoms was also used to treat whooping cough and other dry, irritable coughs.

Red Clover flowers are officially listed in the 1971 edition of the *British Herbal*

CLOVER, RED:
*(Left) flowering stalk,
and (far left) three leaf variations,
all approximately life size.*

Pharmacopoeia. Homeopathic doctors still use the dried flowers to treat various skin diseases. The plant has been recently reported to be antineoplastic (inhibiting tumor growth). While this has not yet been substantiated in mammalian tests, research is being conducted for possible application in modern medicine.

Suggested Uses: The tender young leaves of Red Clover can be added to salads.

CLOVER SYRUP

about 1 quart fresh Red Clover blossoms, 2 cups sugar or honey, 1½ cups water

Crush blossoms gently, then combine all ingredients. Over low heat, bring to the boiling point and simmer for about 15 minutes. Let cool. Strain and bottle. This syrup is soothing for coughs and sore throats and makes a pleasant flavoring for tea.

RED CLOVER TEA

Pour 1 cup of boiling water over 2 tablespoons fresh or dried blossoms. Let steep about 5 minutes, strain, and serve with honey.

To dry Red Clover blossoms: Pick them in the morning after the dew has dried off. Select only the fresh, newly opened flowers, avoiding any that look withered or are tinged with brown. Remove the stems and spread on trays. Do not crowd the blossoms. Allow to dry in an airy place, away from direct sunlight. When thoroughly dry, they will be crisp to the touch. The idea is to retain as much of the color as possible, so store them away from the light, in tightly closed jars.

Dried Red Clover blossoms make a pleasant addition to potpourris and sachets.

CLOVER, WHITE *(Trifolium repens)*

Folknames: Claver, Meadow Trefoil, Garden Clover, Dutch Clover, Clover Grass.

Location: Throughout the city.

Botanical Description: White Clover is a perennial, usually growing only a few inches high, although plants may occasionally be as much as six inches tall. It is of creeping habit. The leaves are divided into three rounded leaflets, sometimes finely toothed at the edges, and have a white spot or band. The actual number of leaflets varies in Clover populations from one (no distinct leaflets) to as many as nine. The flower heads, fragrant and globular in shape, are actually made up of many tiny separate florets. They are white with an occasional tint of delicate pink. In the eastern United States, White Clover blooms profusely from early May until October.

This ubiquitous little plant is an alien. Brought to this country from Europe, where it was highly esteemed, it is now widespread throughout the United States. Often planted as a soil improver and cover crop, Clover enriches impoverished soil because its roots contain nitrogen-fixing bacteria.

Historical Lore, Legends, and Uses: The word "clover" is derived from the Latin *clava,* meaning "club" or "cudgel." This name is attributed to the resemblance the leaflets bore to the famous Club of Hercules, which had three prominent knobs. This ancient association has been preserved down to the present, and is the origin of the suit of clubs in the playing-card deck.

The Anglo-Saxons called Clover *Claverwort,* a word modified during the Middle Ages to *Claver,* a popular name in Tudor England. Clover grew so profusely in parts of Britain that it became incorporated into the place names of several areas, of which Clavering in Essex and Claverton in Cheshire are only two examples.

White Clover is a valued soil improver, bee plant, and forage crop. Cattle were so fond of eating it that one early writer cautioned: "Clover is so powerful to fatten cattle that they must be stopped from eating too much of it, lest they grow so fat that suffocation should ensue."

Bees will work White Clover longer than any other flower, and from this source produce honey of a superior flavor. As a matter of fact, the United States is the world's leading supplier of the honey. This honey is of the best quality and nearly all of it is derived from White Clover.

Tenth-century Anglo-Saxon leechbooks described the following remedy: "For a sore on the right side: work thyme and radish and white clover to a paste and to a drink." For a sore on the left side, there were different instructions: "Pound woodruff in vinegar and work it to a paste, bind it on the left side." We are not enlightened as to why there are two different preparations for the same problem.

The leechbooks also recommended White Clover as a charm against diseases of the mouth: "For sore jaws and the roof of the mouth, if one has with him a root of this Wort and wear it around his neck, this will not trouble him."

Medieval herbalists prescribed White Clover to treat a wide variety of afflictions ranging from falling hair, venomous bites and assorted fevers, to the plague. Clover was also popularly used to cure dizziness and headache. For this purpose, the flowers were boiled in water, honey, and wine. Compresses were dipped in this liquid and applied to the forehead.

Magical properties have long been attributed to White Clover. The Druids

CLOVER, WHITE: (*Above*) *flowering stem,*
and (*right*) *four-leaf variation,*
both approximately life size.

worshiped the plant and considered it sacred—it represented the Great Triad of earth, sea, and heaven. Much later, Clover came to be honored by the Church; St. Patrick taught the doctrine of the Trinity from it. The trefoil motif was popular with artists and became one of the most widely employed ornamental devices.

Three-leafed Clover, an herb of Venus, was commonly used in love matters. To dream of a field of Clover before a wedding foretold a prosperous, joyful married life. Clover brought happiness and good fortune if worn on the person or kept in the house. It warded off enchantments and spells and had the power to protect people and animals against the evil machinations of witches. An old rhyme attests to this belief:

> *Trefoil, Vervain, St. John's Wort, Dill,*
> *Hinder witches of their will.*

The two-leafed Clover was considered a love charm in Essex. A youth or maiden who found one (it was even rarer than the four-leafed Clover) wore it and recited:

> *A clover, a clover, a clover of two,*
> *Put it in your right shoe.*
> *The first young man [or woman] you meet,*
> *You shall have him, or one of his name.*

The five-leafed Clover was rarest of all. It was unlucky to keep, and if you found one, you had to give it away immediately or suffer grave consequences. It would, however, bring good luck to the receiver.

The four-leafed Clover, as everyone knows, has been a prized good-luck charm for centuries. It was formerly supposed that whoever found one would, in addition to having good fortune, be able to discover witches, detect evil spirits, and see fairies. A Cornish legend relates that a milkmaid, after she finished milking the cows one evening, gathered a few handfuls of grass and Clover as a cushion for her head, in order to balance the heavy milkpail and make it easier to carry.

As she walked back through the meadows, the girl was amazed to see bands of fairies and tiny sprites dancing all around her. She pointed them out to others, but no one else could see them. After she arrived home, she discovered the source of her vision: caught up in the grasses she had plucked was a four-leafed Clover.

Two- and five-leafed
White Clover.

The belief in the luck of the four-leafed Clover has by no means been discarded. A popular song has even been written about it, and "genuine" four-leafed Clovers (usually embedded in plastic) are still made up into jewelry or attached to key chains. As far as I know, the wearer of today does not expect to "discover witches," but luck and good fortune are as fervently wished for as ever.

The leaves of White Clover are still used by country people as a weather sign; they are said to feel rough to the touch as a storm approaches, signifying rain.

Herbal doctors still employ preparations of White Clover to ward off mumps.

Suggested Uses: White Clover flowers can be dried and added to potpourri mixtures, and the young leaves can be added sparingly to green salads.

DAISY *(Chrysanthemum leucanthemum)*

Folknames: Field Daisy, Oxeye Daisy, Butter Daisy, Moon Daisy, Dun Daisy, Maudlin Daisy, Maudlinwort, Marguerite.

Location: Road embankments, railroad rights-of-way, vacant lots, open areas.

Botanical Description: Daisies grow about one or two feet tall. The stems are wiry and stiff and occasionally forked. The leaves are small, oblong, and irregularly toothed.

The familiar Daisy flower head with its yellow center and white petals is actually composed of two separate and distinct flowers. The white "petals" are ray flowers, containing pistils (pollen-bearing parts) only; the yellow center, or disk flower, is composed of hundreds of tiny starlike flowers containing both pistils and stamens (male and female parts). These distinctions are easily observed with an inexpensive hand lens. Children will particularly enjoy this.

In the eastern half of the United States, Daisies start to bloom in May and are most profuse during the month of June. Occasional flowers may be seen in July and August. Daisies often form large clumps and when in flower completely transform the landscape (in the city this is usually the grassy sides of a traffic-choked highway) with their beauty.

Daisies are not native. They were introduced from Europe and have become naturalized and are widespread throughout the United States. They are perennials and reproduce from seeds.

DAISY: *Flowering stalk,*
approximately life size.

Historical Lore, Legends, and Uses: During classical times, the Daisy was considered a plant of the moon and was sacred to the goddess Artemis. She was believed able to ward off lightning by interceding with her father, Zeus, god of Thunder. This association was preserved in folknames such as Thunder Daisy, Dun Daisy ("dun" is derived from an old Scandinavian word meaning "boom"), and Moon Daisy. Much later, the Catholic Church dedicated the plant to St. Margaret and St. Mary Magdalene. Names such as Marguerites, Maudeline, and Maudlinwort were common names for Daisies in medieval times. The flowers were particularly beloved by Chaucer. He called them the "ee of the daie" and "day's eye"—it is from these appropriately poetic names that our own word "daisy" is derived.

The Daisy was long esteemed as a medicinal plant. It was a specific for women's ailments, again because of its association with the goddess Artemis, who was the special protector of women. According to the old writers, it was "a wound herb of good repute, seldom left out of those drinks and salves that are for wounds." The leaves were made into an ointment with wax, oil, and turpentine. This salve was considered "most excellent for running sores that were slow to heal, such as in the knee, elbow, and other joints." (These sores must have been rather common in past centuries. Opportunities for bathing were few, and garments were often made from fabrics that were heavy and coarse. Their stiff folds and creases must have rubbed and irritated the skin most painfully. Sores or infections were probably common, and an ointment to heal them must have been gratefully welcomed.)

The juice of the Daisy plant was used for ruptures, "inward burnings," and inflammations generally. An old remedy for broken ribs was to drink this juice and to apply a poultice of the juice mixed with sweet milk and wheaten flour. The flowers were used to prepare a soothing tea for coughs, wheezing, and other chest disorders. The leaves were applied to inflamed eyes and used to heal burns.

The Daisy was one of St. John's sacred plants, always included in bouquets and garlands used for Midsummer Eve (St. John's Eve) festivities. Daisies gathered on St. John's Eve had special power. They were always picked and never pulled up by the root, for to do so even accidentally would stunt the growth of children of the house. According to an old superstition, the roots of Daisies boiled in milk and fed to puppies would stunt their growth and keep them little.

For love divining, a maiden was blindfolded and led to a clump of Daisies. She would reach out her hand and pick some, and the number of flowers she picked would foretell how many years she would have to wait for a husband. In Sweden on St. John's Eve, young girls fashioned a bouquet composed of nine different flowers from nine different places. Daisies were always included. If this bouquet was put under the pillow at night, a girl's future husband would appear in a dream.

To this very day, there are among us those who cannot see a Daisy without picking it and plucking the petals one by one, reciting an ancient charm: "He loves me, he loves me not . . . ," fervently hoping (because we are still superstitious) the last petal will be "he loves me."

Suggested Uses: Daisy leaves are edible and nutritious. They can be chopped and added to salads, but this should be done sparingly, for they can be somewhat bitter. Try to gather very tender young leaves. The white petals can be used in salads too.

A handful of fresh Daisies are delightful to add to a bath, and when dried, they may be used to add color to potpourri mixtures.

Daisy wine can be made in the same manner as Dandelion wine (page 55), but 4 quarts of fresh Daisy flowers should be used. Daisy cordial can be made also, using the same amount of Daisies as Dandelions (see page 55).

DANDELION *(Taraxacum officinale)*

Folknames: Puffball, Blowball, Clockflower, Telltime, Bitterwort, Swine's Snout, Priest's Crown, Piss-in-Bed, Pissenlit.

Location: Everywhere, even in the cracks of roads and sidewalks.

Botanical Description: Who has not seen the Dandelion? It is so common that it seems hardly in need of description. When most people think of a "weed" it is the Dandelion that most often comes to mind, and this is too bad, because Dandelions are really quite cheerful little plants, especially in the city.

Dandelion leaves grow in a rosette from a long tap root. They are dark green, long, and narrow, with edges that are usually but not necessarily coarsely toothed. Hollow flower stems grow from the center of the rosette and ooze a bitter, milky white juice when picked.

In the Northeast the bright yellow flower heads appear early in March. They are most profuse during the months of April and May, but have been seen as late as November. Dandelion "flowers" are actually hundreds of individual, tiny strap-shaped florets combined into a single flower head, and are important pollen sources for honeybees. Extremely sensitive to sunlight, Dandelions open early on sunny days and close at dusk. They react to weather conditions, too, and will close early in the day if rain threatens.

Dandelion is a perennial and reproduces from seeds. The fluffy round seed heads are especially beloved by small children, who love to blow on them and recite all sorts of rhymes and chants—alluded to in such folknames as Clockflower, Blowball, and Telltime—as they watch the little "parachutes" sail away on the wind. These very seeds are perhaps the secret of this plant's biological success; the germination of 90 percent of the seeds of a single plant is not unusual.

Dandelions have a taproot that is almost impossible to remove from the ground in one piece, and any broken-off bits will produce a new plant. This taproot grows straight down and has been known to reach a length of three feet or more. It has a useful purpose as far as nature (if not man) is concerned, for it transports minerals, especially calcium, up to the surface of the soil from deeper levels, enriching it for the benefit of more shallow-rooted plants. After Dandelions die, the root channels left in the soil are helpful to earthworms.

There is some controversy among specialists as to whether Dandelions are indigenous to the New World or were introduced here from Europe. In any event, they now grow everywhere in the United States.

I have always admired the Dandelion's ability to survive. The most ruthless pruning (as with a frequently mowed lawn) will not kill it, as would be the case with most flowering plants. Reduced in size, perhaps to only an inch in height and with correspondingly smaller flowers, the Dandelion will not only survive but produce viable seed.

Gardeners, who are usually gentle souls tolerant of most things green, will often attack Dandelions with a most uncharacteristic viciousness. Continuous warfare is waged against the plants with an arsenal of elaborate digging tools, powerful herbicides (some have been developed that will kill only Dandelions), and even flamethrowers. In spite of these vigorous attempts at extermination, however, Dandelions are holding their own quite nicely.

If it weren't for the Dandelion's cheerful flowers, it might be difficult to tell that spring had arrived in the city. In the most dreary, inhospitable places, where cement covers everything and the dingy browns and grays are made even more drab by dust and pollution, a Dandelion will have found a crack to grow in and, covering itself with flowers, pierce the gloom like so many little suns.

Historical Lore, Legends, and Uses: The origin of the name "Dandelion" has been the subject of much speculation; it is said to have been derived from the French *dent de lion,* "lion's tooth." The jagged, pointed edges of the leaves—even though this is highly variable from plant to plant—are supposed to resemble the individual teeth in a lion's mouth or the row of teeth in his jaw, and with minor variations, this theory (admittedly far-fetched) has nevertheless been generally accepted.

I myself have another, perhaps more plausible explanation. The Dandelion is a plant of the sun and has been associated with the sun since very ancient times. Its color, disklike shape, and raylike florets resemble the sun in miniature, and the rising and setting of the sun influence the opening and closing of the flowers. The lion has, since antiquity, been the animal symbol of the sun. The sun was called *helios* in Greek, and the Greek word for lion is *leon,* or *leo.* The Dandelion was known in Latin as *dens leonus,* and *dens* does mean tooth. But I believe the real connection between the lion and this plant was made because the flower was influenced by and even resembled the sun and the sun was symbolized by the lion.

The explanations of the common folknames for this plant are quite imaginative, even ribald. Dandelion had a reputation as a powerful diuretic, accounting for names such as Pissenlit and Piss-in-Bed. There was even an old saying popular with English children: if they noticed another child picking the flowers, they would teasingly chant, "Pick a Dandelion and wet the bed!" The blunt tips of the unopened buds suggested to some countryfolk the shape of a pig's nose, hence the name Swine's Snout. The round white disk left on the stem after the seeds were dispersed was surrounded with a drooping fringe, reminiscent of the shaven tonsure of a priest or monk. This accounts for the medieval name Priest's Crown.

According to one account, Dandelions were brought to the New World by traders with the Hudson's Bay Company at Fort Churchill (now part of Canada) to augment an unwholesome, mostly meat diet. It has also been said that during pioneer days, women from the East who were homesick on the vast prairies sought comfort by raising a tiny plot of Dandelions to remind them of home. The plants were fenced off to keep out gophers, and watered from the family's precious supply.

In the past, Dandelion was considered a most valuable medicinal plant. The leaves and roots were supposed to be a powerful diuretic and were specific for treating urinary and liver disorders. William Coles, a seventeenth-century physician-herbalist, stated: "It is very effectual for the obstructions of the liver, gall and spleen, and the distempers that arise therefrom, jaundice, and the hypochondriacal passion. It wonderfully opens the urinary tract." The same writer further observed that "children given this remedy produced such great quantities of urine that they wet their beds at night."

Dandelion was credited with many of the same properties as Chicory, and was often substituted for it. A broth or wine made with Dandelion was considered excellent for strengthening the system, especially if taken over a long period of time, and procured sleep and rest for persons exhausted by fevers.

The Chinese are well acquainted with Dandelion. It grows throughout the coun-

DANDELION: *Flowering plant,
approximately life size.*

try, especially in the Yangtze River valley. Some of their more picturesque names for it are Flowering-and-Hoeing-Weed, Yellow-Flowered-Earth Nail, and Golden Hairpin Weed. The entire plant is believed to have tonic properties, and Chinese herbal doctors prescribe it for the relief of abscesses, swellings, and snakebite. The tender young shoots are eaten as a vegetable.

In our own country, the Mohegan Indians of Connecticut used the roots to make a tonic, and a strong infusion of the leaves was administered as a physic. Dandelion juice was once a major ingredient of patent medicines, and until quite recently was listed in the official American and British pharmacopoeias. It is still employed in modern herbal medicine. For this purpose the roots are collected, preferably in the fall, and are used fresh or dried. Dandelion has even been used cosmetically. The flowers, picked just as they start to open, made a fine complexion wash, and the milky juice of the stems was supposed to remove freckles. A poultice of the crushed plant was thought to be good for skin rashes. Old herbalists claimed that the juice "caused new haire to grow" and "laid down the hairs of the eyebrows."

Dandelion is a plant steeped in folklore and was even a popular love oracle. If you were separated from your beloved, you very carefully plucked a full seed head and whispered a message to it. Then, facing where your lover was, you blew hard at the fluffy globe. If you succeeded in making all the seeds fly away at once, your message would be received. A variant in some regions was that if one seed remained, your lover was thinking of you. To dream of Dandelions was a bad omen, however. It foretold misfortune, and deceit on the part of a loved one. Dandelion is said to be one of the "bitter herbs" (traditionally there were five) mentioned in the Book of Exodus to be eaten with the paschal lamb.

There are several legends about Dandelions. One, a story children love, is about how the Dandelion came to be:

Centuries ago, an old miser found a pot of gold at the end of the rainbow. Instead of sharing it, he decided to bury it in the ground, so that only he could find it. He took the gold home, put it into a sack, and went to bed. While he was asleep, however, a mouse searching for food gnawed a hole in the sack.

The next morning, very early so that none would see or follow him, the old man got up and went off to the deep woods to bury his treasure. He was so intent on his plan that he did not notice the coins dropping out of the sack one by one through the hole the mouse had made. Finally, when he reached the darkest part of the forest, he realized the sack seemed very light. He looked inside, and sure enough, it was empty, except for a few coins that were left at the bottom. "Aha," he cried. "All my gold has fallen out! Well, I'll just go back and pick it all up again!" He looked around. "That should be very easy, for I can see them from here, shining in the grass."

He hurried back and bent down to pick them up. But to his astonishment, he could not, for the gold pieces seemed rooted to the ground! Sure enough, when the old miser looked closely, he saw that what had been a shining golden coin was now a beautiful yellow flower! Amazed, he hurried from one golden spot to another, but all the coins had taken root; for the wood sprites had overheard his plan to hide the gold away, and to punish him for his selfishness, they changed the coins into Dandelion flowers, for everyone to love and share.

The Cherokees have their own pretty legend about the Dandelion: one day, the South Wind saw, far off in a distant meadow, a beautiful golden-haired maiden. He was greatly taken with her beauty, and longed to woo her. But he was extremely lazy and somehow never did so. Much later, he thought he saw her again, but her golden hair had turned white. He immediately suspected that his brother the North Wind had tricked him, or stolen her away, and in his anger he blew into the air with all his might. Light as snowflakes, the fluffy seeds scattered; she was gone. The South Wind mourned for her, but to this day he does not know that it was the Dandelion and not a fair-haired maiden that he saw so long ago.

Suggested Uses: Dandelions make good eating, and are good for you too. There are 14,000 international units (IU) vitamin A and 35 milligrams vitamin C per 100-gram serving; the recommended daily allowance is 5,000 IU vitamin A and about 75 milligrams vitamin C. Among the Pennsylvania Dutch it was customary to put a bowl of Dandelion greens on the table on Maundy (Green) Thursday, for if you ate Dandelion on that day, you would stay healthy for the rest of the year.

The leaves are splendid in salad, a fact long recognized and appreciated by the French and Italians.

Dandelion is grown commercially. Large, expensive bunches of it are sold in gourmet and ethnic greengrocers, and home gardeners can obtain seeds to grow their own. The leaves of the wild Dandelion, which taste just like the cultivated variety, can be gathered at one's very feet for free. Just be sure to pick leaves from young plants that have not yet flowered, or they will be bitter.

The vegetable coffees found in health food stores often contain Dandelion roots that have been dried, roasted, and ground. A well-known substitute for real coffee, it is sometimes preferred by people who do not care for coffee's objectionable side effects. Believed beneficial to the system and an aid to digestion, Dandelion roots are said to have a flavor similar to genuine coffee.

Dandelion wine was a beloved homemade beverage in our country and is still prepared in the old-fashioned way in many areas. I have not personally tried this recipe, but it is quite old and reputed to be a fine one.

DANDELION WINE

Makes approximately 3 quarts.

3 quarts Dandelion flowers, 2 lemons, 1 orange, 3 pounds white sugar,
1 package active dry yeast

Remove the green parts of the Dandelion flowers and place the flowers in a large, clean crock or jar. Bring 4 quarts of water to a boil and pour it over the flowers. Let stand three days, stirring once a day. On the fourth day, squeeze the lemons and orange and set the juice aside, then add the sugar and the rinds of the lemons and orange to the flowers and water. Bring all ingredients to a full boil and simmer 1 hour. Add the citrus juices, let cool, and pour all back into the jar. Soften the yeast in ¼ cup of warm water, spread this paste on a slice of toast, and float it on top of the liquid. Let stand three more days. Now strain the liquid, pour it back into the jar, and let stand overnight. Next day, strain through filter paper and pour into clean, sterilized bottles. Cork lightly until the bubbling (fermentation) stops, then secure tops firmly. Let the wine mature at least six months before drinking.

DANDELION CORDIAL

This cordial is not syrupy-sweet. It has a pleasant, unusual flavor and a fine, dark-yellow color. It is nice by itself or on ice with a slice of lemon. As a toddy with hot water and honey, it is soothing for colds and coughs or just pleasant and warming on a cold winter night. This is my own personal recipe, and it is very simple to make. Makes approximately 1 quart.

2 or 3 cups Dandelion flowers, ⅔ cup sugar, rind of ½ lemon, 1 quart vodka

Cut off green bottoms (bracts and receptacles) of the Dandelion flowers, but do not wash them. Mix all the ingredients together and pour into a jar. Cap tightly and put away in a dark place. Shake every day to make sure sugar dissolves. Let stand two weeks, then strain through filter paper. Store in a bottle with a tight-fitting cap.

MIXED GREENS, SOUTHERN STYLE

Dandelion greens can be cooked or added raw to salads. Pick the leaves from plants that have not yet flowered; these are the least bitter. Older leaves from flowering plants can be precooked for 5 minutes in boiling water to remove bitterness, then used in the following recipe. Makes four to six servings.

1 bunch each Mustard, Chicory, and Dandelion greens, ½ pound bacon or ham,
1 or 2 red pepper pods, 1 tablespoon cider vinegar, salt and freshly ground pepper,
chopped scallions or onions

Wash greens well and cut off any tough stems. Put into a colander and pour boiling water over them. Drain and place them into a pot with the meat and pepper pods, and pour fresh boiling water over them to barely cover. Simmer about 30 minutes, then add vinegar, and salt and pepper to taste. Remove the cover so that cooking liquid can reduce somewhat, and simmer 15 more minutes. Drain, reserving liquid.

Place greens in the middle of a platter and arrange the meat around it. Garnish with the chopped scallions or onions. Serve with homemade cornbread and cups of the "pot liquor" to dip the cornbread into.

Dandelion greens can be substituted for Chicory in recipes calling for Chicory leaves (see pages 38–9).

DANDELION SKIN MASQUE

Simmer 1 large handful freshly opened Dandelion flowers in 2 cups water for 15 minutes and let cool until tepid. Then strain and apply the liquid to the skin. Let it dry, then rinse off with lukewarm water. This masque tones the skin and soothes minor irritations.

Dandelions are a source of natural dye. Light yellow can be obtained from the flowers. See the Appendix, page 171, for instructions on preparing dye.

DOCK *(Rumex crispus)*

Folknames: Yellow Dock, Curly Dock.

Location: Roadsides, vacant lots, waste areas.

Botanical Description: Dock is a big plant. The thick stalks (there may be several or just one) are sometimes three or four feet tall. The leaves are large, between six and ten inches long. They are rather narrow and pointed at the tips; the edges are curled or wavy. Flowers on spikes at least twelve inches tall are produced from June through September in the eastern half of the country. The individual flowers are minute, green, and delicate, in dense clusters on the stems.

As fall approaches, the seed spikes become an attractive and distinctive rust brown, and have probably been seen by everyone in the city at one time or another. These seed spikes often persist throughout the winter and make excellent food for wild birds.

The individual flowers and seeds of Dock are worth observing through a magnifying glass. Each tiny flower dangles on a short stalk as thin and fragile as a thread. As the seed develops, the calyx, resembling a frilly miniature bonnet, encloses the seed much as an old-fashioned bonnet enclosed a person's head.

There are several other varieties of Dock. One of them, round-leafed Dock (*Rumex obtusifolius*), is somewhat less common, but frequently encountered in the city. The leaves are oval in shape, wider, and rounded at the tips.

Docks are not native plants. They were introduced from Europe and are now widely naturalized throughout the United States. They are very difficult to eradicate where unwanted. Dock is a perennial with a stout taproot that is hard to remove, and broken-off pieces left in the ground quickly sprout new plants. A large plant can produce more than 30,000 seeds, which are capable of remaining dormant in the soil for fifty years, awaiting suitable growing conditions.

Historical Lore, Legends, and Uses: With the rise of agriculture, Dock, once esteemed, fell into disrepute, becoming characterized as a "noxious plant" or "pernicious weed." It has the ability—which it shares with other plants such as Dandelion and Ragweed—to survive and set seeds under conditions that would destroy most other plants; this accounts for its success and prevalence in an urban environment. If Dock is frequently cut, mowed, or otherwise deprived, it will nevertheless grow into a complete miniature plant. I have seen Dock plants no more than six inches high, with leaves three or four inches long as opposed to the usual six or ten inches, and a seed stalk that had managed to produce (I counted them) over one hundred seeds.

Dr. Thomas Green, a nineteenth-century English botanist, describes an experience illustrating the tenacity of Dock. The story was related to him by one Dr. Marshall. This gentleman described having seen a thick bed of Dock plants, greedily eaten by pigs in addition to being mowed several times over a period of years, until in Dr. Marshall's words, "it vanished as if by a charm." Dr. Green retorted that "neither swine nor scythe could have killed these plants, which evidently died of old age." Dr. Green despised Dock. He considered it to be troublesome and injurious to agriculture, a ruiner of pastureland and an especially severe pest of grasses. No mincer of words, Dr. Green stated flatly that, in his opinion, to permit Dock which had ripened its seeds to be threshed with Clover (planted by farmers as a forage crop and soil builder) was an unpardonable neglect amounting to an actual crime.

Dock was considered an important medicinal plant as far back as classical Greece. Docks were formerly members of a genus called *Lapathum,* a name derived from a Greek word meaning "to cleanse," an allusion to the reputation of this herb as a diuretic, laxative, and blood purifier. It was especially valuable for treating diseases of the blood and liver. The root was believed the strongest medicinal part, but the leaves and seeds were employed as well.

Dock was highly esteemed by the Anglo-Saxons. To reduce swollen glands in the groin, a mixture of Dock leaves and grease was wrapped in a cabbage leaf, warmed in hot ashes, and applied as a plaster. For an alarming condition described only as a "worm-eaten and mortified body" a tenth-century Anglo-Saxon leechbook prescribed a salve of Dock to "smear therewith at night." It does not say whether or not this helped.

Dr. William Coles, a seventeenth-century herbalist, described two kinds of Dock. One had a yellow root—this he called simply Dock; the other had a red root and he called it Bloodwort: "The reddish and yellowish color of the insides of the roots of divers Docks do signify that they are good for hot livers and jaundice," he counseled. "The red in them represents the liver, the yellow the jaundice, and therefore Bloodwort is most effective for the first and Dock for Jaundice." The leaves of all varieties were supposed to "soften the belly" (that is, relieve constipation). For this purpose, the leaves were chopped, boiled, and eaten. Dock seeds were good to be drunk in wine or water "against the loathesomeness of the stomach." The roots infused in wine "assuaged the pain in the teeth," and this same liquid would cure itching if the troubled area was bathed with it.

In America a Vermont physician who practiced during the early part of the nineteenth century praised Dock highly: "Though it is despised by many as a cumbersome weed, it is one of the finest productions of which our country can boast. It is good for coughs, asthma and pleurisy, and an infusion of the seeds in Rum helps cure disorders of the breast and lungs." We are at a loss to understand what the doctor had in mind when he stated that "the leaves [of Dock] applied to the soles of the feet make a revulsion from the head."

The Delaware Indians employed Dock as a blood conditioner and jaundice remedy. Several other native American peoples, especially those of the Southeast and the Central Plains, used the crushed leaves and roots to treat cuts and boils. These practices may have been learned from the early settlers, since Dock was not a native American species, and its use for these ailments was well known in Europe. The Pennsylvania Dutch knew Dock as *Halwer Gaul* and made a tea from the root. It was drunk as a tonic and used to treat liver ailments.

Dock, a popular ingredient in patent medicines, probably did less harm than

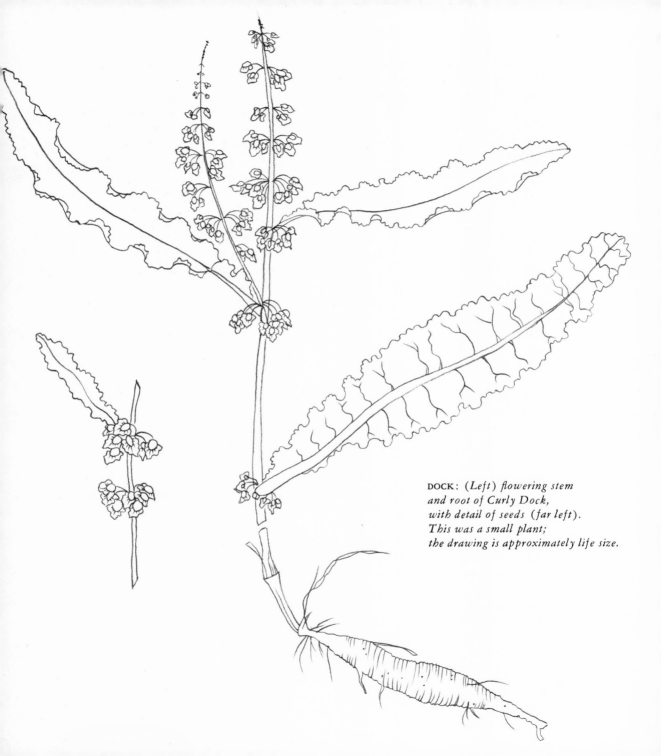

DOCK: (*Left*) *flowering stem
and root of Curly Dock,
with detail of seeds* (*far left*).
*This was a small plant;
the drawing is approximately life size.*

the other ingredients, which were usually opium or grain alcohol.

At one time, Dock was the main ingredient in an old-time "cancer balm." This "balm" was credited with remarkable cures, but it is hardly possible that it cured actual malignancies. It is likely that harmless (and temporary) lumps and miscellaneous swellings were mistaken by backcountry "doctors" for actual cancer.

Dock is listed in the 1971 edition of the *British Herbal Pharmacopoeia*. Homeopaths use the fresh or dried root. It is considered valuable as a gentle purge and useful in certain chronic skin diseases, and is often combined with Dandelion.

Dock is well known in China and valued by the Chinese for the same properties attributed to it in Europe and America. The root is used to treat a variety of skin disorders and the seeds to treat dysentery.

Folklore and superstition have long been associated with Dock. One of the oldest and best-known examples of the latter was its use as a cure for nettle rash. It was a common country belief that Dock would always be found growing near nettle (*Urtica dioica*), a plant covered with fine, stinging hairs that cause painful inflammation if touched. If this happened, Dock leaves were plucked, crushed, and laid on the afflicted area, accompanied by the following words:

> *Nettle in, Dock out,*
> *Dock in, Nettle out,*
> *Nettle in, Dock out,*
> *Dock rub nettle out.*

Children recited a similar rhyme:

> *Out Nettle, in Dock,*
> *Dock shall have a new Smock,*
> *But Nettle shan't have nothing!*

The use of Dock to cure nettle rash is met with so frequently that there may well be something to it. It is quite possible that a poultice of Dock leaves cures the sting caused by handling nettles.

Another very old belief had it that Dock seeds tied to the left side of the body would cause conception, but we are not told whether the man, woman, or both were supposed to do this, nor under what circumstances.

The most unusual and fascinating of all the charms involving Dock is to be found in a tenth-century Anglo-Saxon volume known as *The Leechbook of Bald:* "For one afflicted with Elf Disease" (made ill by witchcraft), Dock and other herbs were chopped and mixed with ale and holy water, and the following charm was sung three times over the potion:

I have wreathed round the wounds
the best of healing worts,
that the baneful sores may
neither burn nor burst,
nor find their way further
nor turn foul and fallow,
nor thump or throb on
nor be wicked wounds,
nor dig deeply down.
But he himself may hold
in a way to health.
Let it ache thee no more
than an ear in earth acheth.

The last line refers to a corpse in the earth who obviously feels no pain. "Sing this many times, that the earth may bear on thee with all her might and main. These charms may a man sing over a wound." The Anglo-Saxons were nominally Christian (indeed, Bald, the compiler of this leechbook, is believed to have been a monk), but this charm is quite pagan and, in addition to the use of the strongly magical number 3, invokes the aid of the powerful earth spirits as well.

An English writer described a charm, partly medical and partly superstitious, extant in Britain as recently as the early 1900s: "Workers in ironstone quarries, who sometimes develop a peculiar sore on the forearm as a result of their work, sometimes cut a Dock root and rub the cut surface on the sore. This is done *every third day* [italics mine] until the cure is complete."

DOCK ALE

This is a seventeenth-century recipe for a medicinal ale reputed to cool and cleanse the blood. It will also cure "all manner of itch and manginess of the whole body, and makes wenches to look fair and cherry-like." I have not tested it.

Take half a pound each of Dock and Madder root, of Senna leaves four ounces, of Anise Seed and Liquorice two ounces each, and one handful each of Scabious and Agrimony. Slice the roots, bruise the seeds and Liquorice, and break up the herbs with your hand. Steep three days in strong ale. Drink this liquid as you ordinarily drink for at least three weeks, though the longer you take it the better.

According to Julia Morton, a contemporary American writer, Dock is a famous southern potherb, even richer in flavor than more well-known greens such as collard and turnip. "Every garden should have a patch of this Dock," she advises, "especially since it requires no care once it has become established. During spring and early summer, a few square feet of these weeds will give solid nourishment."

It is not necessary to plant Dock in your garden, because it grows all over and is free for the picking. Gather the smallest and tenderest leaves, preferably in the spring, but any time before the plant flowers. After that the leaves become bitter.

Suggested Uses: Dock is very high in vitamin A and vitamin C: a 100-gram portion contains 12,900 IU vitamin A and 119 milligrams vitamin C; the recommended daily allowance is 5,000 IU vitamin A and about 75 milligrams vitamin C.

DOCK COUNTRY-STYLE
Dock leaves, meaty ham bone, cider vinegar, sugar, salt (optional)

Parboil the leaves for 10 minutes, then drain. Cover with a fresh change of water, add the ham bone, and simmer gently for 15 to 20 minutes. Drain Dock and keep hot. Reduce the liquid somewhat, shred ham from bone, and return ham to pot. Add vinegar and sugar to taste (it should be sweet-sour), and salt, if needed. Pour this dressing over chopped Dock greens and serve hot.

The tall seed stalks of Dock are attractive in dried plant arrangements. Gather them in the summer, when they are a subtle rose-tinted green; or in the fall, when they turn a dramatic rust brown.

Dock is an excellent source of natural dye. Shades of yellow-gold to tan can be obtained from the roots. See the Appendix, page 171, for instructions on preparing dye.

GOLDENROD *(Solidago species)*

Folknames: Woundwort, Aaron's Rod, Canada Goldenrod, Sweet Goldenrod.

Location: Roadsides; vacant lots; open, sunny areas.

Botanical Description: Of the many varieties of Goldenrod that grow in the city, the most common is probably Canada Goldenrod (*Solidago canadensis*). It is a tall plant, growing from one to four feet high. The stems are densely covered with narrow, sharply toothed leaves, somewhat rough to the touch. In the Northeast, Goldenrod blooms from late July through October. The individual yellow flowers are very small but, massed together in clusters, produce the familiar plumelike, graceful flower heads. Canada Goldenrod is a perennial and reproduces from seeds.

Goldenrods are native to North America, and at least sixty species grow throughout the United States and flourish in a variety of habitats. All have the familiar yellow flowers, but the flowering stems themselves may be branched, plumelike, club-shaped, or flat-topped. Goldenrods are difficult to tell apart, and some can be positively identified only by plant specialists.

Historical Lore, Legends, and Uses: Goldenrod is a much-loved herb. The generic name *Solidago* is derived from Latin and means "I make whole," an allusion to the healing qualities associated with this plant. John Muir, the great American naturalist, paid gallant and poetic tribute to Goldenrod: "The fragrance, color and form of the whole spiritual expression of Goldenrod are hopeful and strength-giving beyond any others I know. A single spike is sufficient to heal unbelief and melancholy."

Woundwort (*Solidago virgaurea*) is the only European species of Goldenrod. The leaves and flowering tops were used medicinally. It was "the sovereign wound-herb of many" and, because of its astringent qualities, useful for treating all kinds of sores and ulcers, especially those of the throat and mouth. A decoction of the leaves was even said to "fasten teeth loose in the gums."

During Elizabethan times, Woundwort was in great demand and large quantities of it were imported. In the early 1600s it was discovered growing in England; seen thus as a common plant, Londoners (who were great users of Goldenrod) quickly lost interest in it. This made an unpleasant impression on the herbalist John Gerard. "This is an excellent example of humanity's fecklessness and inconstancy," he observed. "A thing is only held valuable if it is rare and strange."

In 1635, John Parkinson described a plant brought to him from the New World which he called "Golden Rod of America." It had "long, narrow leaves, somewhat dented about the edges and yellow flowers all on one side" and was recognized as being quite different from the European plant.

The beneficial properties of Goldenrod were well known to native American peoples. The Cherokees prepared a tea from one species to reduce fever; other species were employed to treat bladder and kidney ailments. The Delaware used a related species, Early Goldenrod (*Solidago juncea*), as a diarrhea remedy, and for fevers the fresh green leaves were chewed raw or infused in hot water.

Dr. J. Monroe, a nineteenth-century American physician, gave high praise to Goldenrod: "A decoction of this herb is excellent for general weakness and disability. It is a great cleanser of the internal viscera and prevents consumption and dropsy." He adds helpfully, "There are two kinds; the largest is the strongest."

The Pennsylvania Dutch appreciated Goldenrod, as can be seen from the names they had for it: *Feewar Blumma* and *Wondergraut,* as the plant was called, translate respectively as "Fever Bloom" and "Wonder Herb." It was considered good for digestive troubles and, because it induced perspiration, useful in reducing fevers. It was taken as a tea. Pennsylvania Dutch women obtained a yellow dye from the flowers.

The best-loved of all the American Goldenrods was the variety known as Sweet Goldenrod (*Solidago odora*). It is found throughout the Northeast, especially in dry, open woods, but I have found it growing in the city as well. It is about one to three feet tall, and the plumelike yellow flower stalk is smaller and not as showy as Canada Goldenrod. The long, narrow leaves have smooth, unnotched edges and emit a characteristic sweet, aniselike odor when bruised or crushed. In addition to being considered valuable for relieving nausea and stomach spasms, it has a delightful aroma and taste and was one of the best native "tea herbs." During the nineteenth century it was even exported to China.

Sweet Goldenrod had a small but undeniably significant role in American history. After the Boston Tea Party, the patriots were hard put to survive without their favorite beverage. Yankee ingenuity quickly devised a fine and widely accepted substitute. It was called, appropriately, Liberty Tea, and its major ingredient was Sweet Goldenrod. Who knows what the consequences might have been if not for this timely (and tasty) substitute?

Years ago, country people used powdered Goldenrod flowers to make people sneeze. This powder was surreptitiously administered and was considered a fine joke. Of course, any powdered substance introduced into the nostrils will provoke sneezing, but the particular use of Goldenrod in this instance may have planted in people's minds the idea that Goldenrod, rather than the way it was administered, caused sneezing.

While it is certainly possible for someone to be allergic to this plant, this unfortunate association has dogged the inoffensive Goldenrod ever since, and hayfever victims hold Goldenrod partly responsible for their sufferings. Goldenrod blooms profusely at about the same time that hayfever sufferers experience their worst discomfort, but Goldenrod actually sheds insignificant amounts of pollen. The real culprit is Ragweed. It flowers at approximately the same time as Goldenrod, and though it is much less noticeable, it is far more abundant.

GOLDENROD:
Upper part of flowering stalk,
approximately life size.

In England herbal doctors presently employ European Goldenrod (*S. virgaurea*) to relieve indigestion, and a spray or gargle is used for treating nose and throat infections. In France the flowering tips, gathered just as they start to open, are dried and made into a syrup or used as a tea.

In China, Goldenrod has been used as in Europe, to treat wounds and hemorrhages. There is a legend about how its use was discovered: long ago, a man named Chi-nu was in the forest, cutting down a ti plant. He saw a large snake and, because he was frightened, shot it with his bow and arrow. Returning to the spot the next day, he was startled to hear the rhythmic thumping of what sounded like a mortar and pestle. Chi-nu went looking for the source of the noise, and eventually found himself in a hazel thicket where several young men dressed in green tunics were busy crushing Goldenrod plants with a huge mortar and pestle. He asked them why they were so engaged, and they replied that their master had been injured by an arrow and they were preparing medicine that would heal him. They explained the medicinal virtues of Goldenrod to Chi-nu, and he in turn taught it to humanity.

Goldenrod plants are sometimes afflicted by round swellings, caused by an insect that lays its eggs in the stems. These swellings, or galls, were in former years highly prized by New Englanders, who called them rheumaty buds. It was believed that if carried about on the person, the buds would ward off rheumatism for as long as the little grub inside stayed alive.

Suggested Uses: The leaves and flowers of Sweet Goldenrod make a pleasant and healthful beverage, said to be tonic and astringent. Called Blue Mountain Tea, it is stocked by herb shops and health food stores throughout the United States. Use alone or mixed with regular tea in any proportion desired; 1 teaspoon of the leaves to 1 cup of water is a good combination. Pour boiling water over and let steep 10 minutes. Strain and serve with honey and lemon, if desired. Sweet Goldenrod can be gathered in the summer and used fresh or dried. You can identify it by crushing a leaf; the true Sweet Goldenrod will release a distinctive and characteristic odor of anise.

To-dry Sweet Goldenrod leaves: Remove leaves from stems and place them, one layer deep, on flat trays in a room with good air circulation. Do not dry in direct sunlight. When the leaves are thoroughly crisp and dry, store them in jars with tight-fitting lids, away from the sun.

Goldenrod is an excellent source of natural dye. The flowers yield shades of yellow to yellow-green. See the Appendix, page 171, for instructions on preparing dye.

GROUND IVY

(*Glechoma hederacea* or *Nepeta hederacea*)

Folknames: Gill-over-the-ground, Robin-run-in-the-hedge, Lizzy-run-up-the-hedge, Catsfoot, Alehoof, Tunhoof, Field Balm, Creeping Charlie.

Location: Under hedges, along walks, in gardens, waste places, and moist shaded areas generally.

Botanical Description: In spite of its name, Ground Ivy is not an ivy, but a member of the mint family. It resembles ivy closely in growth and habit, which accounts for the name it is so commonly known by.

Ground ivy has long trailing stems that creep along the ground. The leaves are dark green, stalked, and grow opposite each other. They are delicately scalloped at the edges and somewhat downy to the touch. When bruised or crushed, the leaves emit a pleasant aromatic odor reminiscent of mint. The plants bloom from April to June in the Northeast, and the flowers, usually violet in color, grow in whorls at the juncture of leaf and stem.

Ground Ivy is a perennial. The plants set seeds, but reproduction is usually by means of roots that form at the joints where stems and leaves meet. These grow into new plants. As with true ivy, in a sheltered location Ground Ivy can remain green even in cold weather.

An alien originally from Eurasia and introduced from Europe, Ground Ivy has become naturalized and grows throughout the eastern United States.

Historical Lore, Legends, and Uses: The picturesque folknames associated with Ground Ivy refer to its creeping habit (in Latin it was called *corona terrae,* "earth crown," because its foliage was "like a garland over the ground") and its use for flavoring ale. So greatly esteemed was Ground Ivy in brewing ale that variations of the names Alehoof and Tunhoof are found for it in most European languages. Ground Ivy gave ale a good taste, clarified it, prevented souring, and imparted healthful virtues. "Gill" ale was a popular beverage for centuries. During the reign of King Henry VIII, hops, whose cultivation had been prohibited, were finally permitted to be grown in Britain and Ground Ivy, "so common that it was never grown," was quickly supplanted by hops and fell into disuse.

This was not quite the end, however. That respected seventeenth-century English herbalist, Dr. William Coles, observed: "Country people formerly used it

[Ground Ivy] much in their ale and beer, and so they would now, if they were wise. But this age forsakes all old things, though never so good, and embraces all sorts of novelties whatsoever. But," he continues, "the time will come when all the fopperies of the present time shall be slighted, and the true and honest prescriptions of the Ancients come into request again." He rebukes those "so prejudicial to the virtues of simples that they think fabulous whatsoever things are related concerning them, though they stand in great need of them. To try such a thing as this would not require either much time or cost."

Ground Ivy was used medicinally and considered especially beneficial for kidney and lung disorders. A tea was prepared from the leaves, or the leaves were steeped in brandy, and a syrup was made from the whole fresh plant. In addition, the juice dropped in the ears was said to "wonderfully help the noise and singing of them."

The Anglo-Saxons called this plant Earth Ivy and used it to treat headaches. A prescription from a tenth-century leechbook: "That the head may not ache from the heat of the sun, take leaves of this wort, pound them in vinegar and smear on the forehead. This is good for anything that vexes the head."

Ground Ivy has had a long history as a headache cure. The fresh juice squeezed from the leaves was snuffed up the nostrils, and this was a very popular remedy, said to relieve the most stubborn headache.

In the United States, a tea from the leaves was at one time considered to be a remedy for and preventer of a type of lead poisoning known as "painter's colic," but by the nineteenth century, Ground Ivy had fallen into disuse and most physicians doubted that it had any medicinal value at all, though the subject continued to be debated for some years.

Ground Ivy is common in the north and central provinces of China. In that country, most of the folknames for it allude to the resemblance of the leaves to Chinese coins. It was used medicinally to treat toothache and earache, but was believed most valuable in reducing fever.

Modern herbalists feel Ground Ivy soothes inflamed mucous membranes. They prescribe a tea made from the fresh leaves for sore throats, bronchitis, and chest colds.

Ground Ivy was long associated with witchcraft and magic. It was always woven into garlands and crowns worn on Midsummer Eve, a famous night for occult occurrences. Witches were particularly fond of Ground Ivy. They liked to use it in their spells because it grew abundantly in graveyards, ancient ruins, and other appropriately dank and sinister places. This baneful association between witches and Ground Ivy may explain a long-held country belief that Ground Ivy killed any plants growing near it.

GROUND IVY: (*Above*) *leafy stalk,*
and (*left*) *detail of seed stalk,*
approximately life size.

Suggested Uses

GROUND IVY TEA

Ground Ivy has a high vitamin C content. A soothing beverage for colds, coughs, and irritated throats can be prepared by pouring 1 cup of boiling water over 1 or 2 tablespoons of bruised fresh Ground Ivy leaves. Let steep 10 minutes, then strain. Sweeten with honey.

GROUND IVY FACIAL

A facial wash for chapped or sunburned skin can be made by pouring 2 cups of boiling water over 1 cup of chopped fresh Ground Ivy leaves. Cover and let stand until tepid. Strain, pressing out as much liquid as possible, and blot gently on the skin. Rinse off with lukewarm water. This wash is also soothing for tired or sore eyes.

Ground Ivy leaves for tea and facials should always be used fresh rather than dried. The beneficial volatile oils do not survive the drying process.

JAPANESE KNOTWEED
(Polygonum cuspidatum, P. sachalinense)

Folknames: Mexican Bamboo, Japanese Fleeceflower.

Location: Roadsides, embankments, parks, along fences and buildings, vacant lots, and waste areas.

Botanical Description: Japanese Knotweed is a tall, rapidly spreading plant, often more than eight feet high. It has hollow stems, red-spotted when young, then turning entirely red; and large, heart-shaped leaves. In the Northeast during August and September, the gracefully arching stems are covered with drooping clusters of cream-colored or white flowers. The tall, jointed stems persist throughout the winter.

Japanese Knotweed is a perennial and can reproduce from seeds, but it is from long underground runners, spreading many feet from the parent plant in a single season, that Japanese Knotweed forms large thickets with a speed that is truly astonishing. The runners are numerous and very difficult to dig out entirely, and will grow right up through a layer of asphalt. Once this plant has gained a foothold where it is not desired, it is nearly impossible to eradicate.

Japanese Knotweed (or Mexican Bamboo, as it is sometimes inexplicably called, since the plant does not come from Mexico and is not a bamboo) is a member of the buckwheat family and is related to rhubarb and Curly Dock. It was originally introduced from the Orient as an ornamental garden plant in the nineteenth century. It quickly escaped cultivation and became naturalized throughout the northeastern United States. The two species closely resemble each other, with minor differences in the leaf.

Historical Lore, Legends, and Uses: Japanese Knotweed is a relative newcomer to this country. It was intended to be an "exotic" garden plant, and it really is quite lovely, with its red stems, elegant heart-shaped leaves, and foamy white flowers. It was praised in the *Cyclopedia of Horticulture* as "a very effective plant for bold, mass effects." Bold it truly is. Japanese Knotweed escaped cultivation with speed and ease, sometimes pushing out more desirable flora. It is encountered everywhere in the city, but in an urban environment its aggressiveness is a virtue. Japanese Knotweed quickly develops into a beautiful living screen, utterly transforming barren, unsightly areas where it particularly loves to grow.

In China, the root was used medicinally to treat menstrual and postpartum difficulties. It does not seem to have been employed as a medicinal plant in the West.

Most of the literature on Japanese Knotweed is devoted to the best methods for eradicating it, but few people realize its potential as an extremely valuable and versatile food resource. It can be prepared many ways, is free for the gathering, and best of all, tastes truly delicious.

Suggested Uses: Start looking for Japanese Knotweed shoots in the spring. They look very much like asparagus, but have a flavor all their own. Early April is a good time; the thick, red-spotted stalks tipped with tight-rolled leaves are easy to find. Look among the tall, leafless gray stalks of last year's plants. The shoots are best gathered when no more than six or eight inches high.

BOILED JAPANESE KNOTWEED IN THE MANNER OF ASPARAGUS

Gather Japanese Knotweed in the early spring, selecting shoots with the thickest stems. Allow about six per person. Wash well and remove any leaves on the stalks, but leave the tightly rolled leaves at the tips. Place in a pan with about 1 inch of water, or cook in a vegetable steamer. The stems cook quickly; in about 5 minutes they will turn a creamy olive green. Prick with a fork; they will be soft, if done. Do not overcook. Drain and serve hot with melted butter or hollandaise sauce.

JAPANESE KNOTWEED:
(*Right*) *upper stalk
with flowers and seeds,
and* (*far right*) *young shoot,
both approximately life size.*

JAPANESE KNOTWEED PURÉE

The addition of sugar changes the character of Japanese Knotweed entirely. Prepared this way, it is flavorful and pleasantly tart, reminiscent of rhubarb, perhaps, but lacking the sharp, acid taste of rhubarb and requiring far less sugar to make it palatable.

Gather stalks, choosing those with thick stems. Wash well and remove all leaves and tips. Slice stems into 1-inch pieces, put into a pot, and add ¾ cup sugar for every 5 cups of stems. Let stand 20 minutes to extract juices. Add only enough water to keep from scorching, about half a cup. Cook until pieces are soft, adding more water if necessary. They will cook quickly. When done, the Japanese Knotweed needs only to be mixed with a spoon. Add lemon juice to taste and more sugar if desired. Serve chilled for dessert just as it is, or pass a bowl of whipped cream. This purée is excellent spooned over vanilla ice cream or baked in a pie shell. The cooked purée will be olive green in color. If you wish, you may add a few drops of red food coloring to make it rosy. It keeps very well in the refrigerator and may be frozen for later use.

JAPANESE KNOTWEED BREAD

This bread is moist and flavorful. It needs no spices, keeps well, and is excellent toasted and served with butter or cream cheese.

2 cups unbleached flour, ½ cup sugar, 1½ teaspoons baking powder, 1 teaspoon salt,
1 egg, 2 tablespoons salad oil, ¾ cup orange juice, ¾ cup chopped hazelnuts,
1 cup sweetened Japanese Knotweed purée (preceding recipe)

Preheat oven to 350°. Sift dry ingredients together into a large bowl. Beat the egg white with the oil and orange juice. Add along with hazelnuts and purée to dry ingredients. Do not mix until all ingredients are added, and blend only enough to moisten. *Do not overmix.* Spoon gently into buttered 9½-by-5-by-3-inch loaf pan. Bake about 1 hour or until a straw or cake tester inserted in the center comes out dry. Cool by removing from pan and placing it on a rack. For muffins, spoon into buttered muffin tins and bake about 25 minutes. Makes about one dozen.

Japanese Knotweed leaf, much reduced.

LADY'S-THUMB

(*Polygonum persicaria* and related species)

Folknames: Smartweed, Redleg, Red Knees, Pepperplant, Bity-Tongue, Smart-ass, Arsmart, Heartweed, Lover's Pride, Blackheart, Peachwort, Heart's Ease, Virgin's Pinch.

Location: Under hedges, in gardens and parks, along fences and buildings, in sidewalk cracks, vacant lots, and waste areas.

Botanical Description: Lady's-thumb varies from as little as six inches to three feet high. The stems may be erect or sprawling, and have a reddish tint. Several grow from one root. The leaves are smooth-edged and lance-shaped, occasionally marked in the centers with a dark, triangular spot. The joint of the leaf and stem is enlarged and knotlike, and a fringed, papery membrane clasps the stem at each of these joints.

Lady's-thumb blooms from May to November throughout its range. Flowers are borne at the ends of the stems as well as on stalks produced at the juncture of stem and leaf. They are pale pink to rose-colored and clustered tightly together on the stems, resembling miniature clubs. The individual flowers are quite small, and their delicate structure is best appreciated with the help of a hand lens. Lady's-thumb and several other closely related species are very common in the city. A characteristic they share, in addition to the papery sheath at the joints, is a pungent taste varying in sharpness among the species. Wild birds are fond of the seeds.

Lady's-thumb is an alien, introduced from Europe, but several other species are native. An annual that reproduces by seeds, it grows throughout the United States.

Historical Lore, Legends, and Uses: The peppery, biting taste of Lady's-thumb has long been known and accounts for folknames such as Smartweed and Pepper-plant; the more earthy Smartass and Arsmart were popular English names. The Latin name for it was *Persicaria* because its leaves resembled those of the peach, and this appellation was eventually adopted as the scientific name.

Not all *Polygonums* seemed to have the characteristic sharpness, however. One old writer suggested (in somewhat convoluted prose) the following method for distinguishing the peppery-tasting plants from a milder variety, which he called

Dead Arsmart: "Break a leaf across the tongue. The biting sort will make the tongue to smart and so will not the other." Another of his observations was amusingly picturesque: "Lady's-thumb has many knotty joints, called by some men knees, one being always stopped into the other after the manner of a trumpet."

In times past, Lady's-thumb (and various other Smartweeds) were used for a variety of ailments. The whole plant was considered astringent and diuretic. Essences, tinctures, and infusions were prepared and used to treat jaundice and kidney and bladder diseases. A wash made from the fresh leaves steeped in water was considered effective for sores and skin infections, and a tea of the leaves was used as a gargle to promote the healing of mouth and throat sores. A piece of root from the black-spotted variety, bruised and held to an aching tooth, would take away the pain.

The Anglo-Saxons used Lady's-thumb as a remedy for sore eyes and ears. They called it Untrodden to Pieces, perhaps because it was so hardy and tough that it survived even being stepped upon or otherwise crushed.

The plant was capable of healing animals as well as people. It was frequently mentioned that the juice of the herb applied to horse or cattle sores would keep away the flies "that would otherwise stick to them, even in the hottest summer."

Water Pepper (*Polygonum hydropiperoides*), a nearly identical, closely related species, was known to the Pennsylvania Dutch settlers. They called it *Fleeh Graut*. An infusion of the leaves in cold water was used to ease severe cough. When meat was being butchered, Amish farmers burned this herb in pots to keep away flies.

In China, Water Pepper was a potherb as well as a medicinal plant. It was crushed and applied as a poultice to cure snakebite and foot infections. The Chinese used Water Pepper to treat cholera, and by a curious coincidence, in the United States during the terrible cholera epidemics of the early and mid-nineteenth century, patients were wrapped in sheets that had been soaked in a hot decoction of this plant as soon as symptoms of the dread disease appeared.

Lady's-thumb has been used to dye wool yellow. It was formerly much esteemed for this purpose in rural Louisiana, where it was called Culrage. Other varieties have been cultivated in the Orient and are used to dye textiles beautiful shades of green, brown, and blue.

For reasons difficult to explain, Lady's-thumb and similar species of *Polygonum* have been long associated with numerous folk beliefs ranging from the merely colorful to the downright bizarre. An example of the latter is a healing charm from a tenth-century Anglo-Saxon leechbook: "Before sunrise, or shortly before it begins fully to set, go to this wort and scratch it round with a golden ring, and say that thou will take it for a leechdom of the eyes. And after three days go again thereto, before

LADY'S-THUMB: (*Right*) *flowering stalk,*
and (*far right*) *leaf showing dark spot*
and variability of leaf,
both approximately life size.

the rising of the sun, and take it and hang it about a man's neck. It will profit well."

Heart's Ease, a very old name, alludes to the belief that the plant was useful for curing heart ailments, as signified by the heart-shaped spot on the leaves. Others thought the Virgin Mary had pinched the leaves with her thumb and forefinger, and that the dark spot signified that the plant was a remedy for bruises and wounds. Hence the names Lady's-thumb and Virgin's Pinch.

The leaves were strewn in chambers to kill fleas, and according to a superstition usually credited to the Scythians, a handful of Lady's-thumb placed under the saddle of a horse would enable it to travel long distances without thirst, hunger, or fatigue. It is not believed that the Scythians used saddles, but never mind!

William Coles, a seventeenth-century physician and herbalist, recommended the use of Lady's-thumb for the following unhappy condition: "When a woman's belly be great and she not with child, let her boil Arsmart, Rue and Hyssop, of each one handful in a quart of Ale and drink it thereof morning and night. It will reduce it to just measure."

He then adds: "It is said that if a handful of Arsmart is wetted in water and applied to a wound or sore and afterwards buried in moist ground, as the herb rots, so will the sore miraculously heal." It is an odd suggestion. Coles was highly educated, a respected physician, and a good Christian. It is most strange that he seems to endorse pagan belief in the power of sympathetic magic.

According to this philosophy, which is as old as man, "Like caused like." This formula was popular with witches and other workers of harm. They would, for example, rub a wound with the plant or, to avoid suspicion, contrive for another to do so. The result would differ from that described by Coles, however. Having now established a magical connection between the person and the plant, the plant would be buried in moist ground and as the plant rotted, *so would the sore fester,* to the harm or eventual death of the victim.

The following prescription is a similar example: "If a woman drink, fasting, the juice of this plant [Lady's-thumb] three times a day in wine for the space of eleven days after her last show [menses] and the man do the same before he lies with her, she shall conceive and the child shall be male."

This last was recommended in all good faith by William Turner, a renowned Elizabethan physician and a man who knew his herbs. The peculiar thing about it is that Turner was an outspoken skeptic who frequently ridiculed the superstitious practices and fabulous claims of physicians and herbalists of his own and former times.

Most bizarre of all, perhaps, are the virtues the Chinese attributed to a species of Lady's-thumb:

When the root [of this herb] is fifty years old and as large as a fist, it is called *Mountain Slave*. If taken internally for one year, it will preserve the black color of the hair and beard.

When it is one hundred years old it is as large as a bowl and is called *Hill Brother*. If taken internally for one year, it will guarantee a rosy, cheerful countenance.

When the root is one hundred and fifty years old it is as large as a basin and is called *Hill Uncle*. If taken internally for one year, the teeth will fall out and healthy new teeth will grow in their place.

At two hundred years old, the root is the size of a one-peck ozier [*sic*] basket and is called *Hill Father*. If taken internally for one year, the countenance of the elderly will become as that of a youth's and the aged will be able to walk as easily as a young man runs.

At three hundred years of age the root is as large as a three-peck ozier basket and is called *Mountain Spirit*. This has a pure, ethereal substance and if taken internally for a long time one will become an earthly immortal.

Suggested Use: Lady's-thumb is not poisonous, but the leaves have a sharp, peppery taste and are not good to eat. However, a beautiful green dye can be prepared from the whole plant (see page 172).

LAMB'S-QUARTERS *(Chenopodium album)*

Folknames: Goosefoot, Wild Spinach, Allgood, Pigweed, Dirtweed, Dirty Dick, Midden Myles, Baconweed, Fat Hen, Frostbite, White Goosefoot.

Location: Along roadsides, under hedges, in vacant lots and waste areas.

Botanical Description: Lamb's-quarters vary in size from just a few inches high to as much as three or four feet tall. The stems are erect, many-branched, and frequently tinged with red. The leaves are oval or triangular in shape, with toothed edges, and are dark green above with mealy white undersides. This last is a characteristic useful in identifying the plant.

Minute green flowers are borne on stalks growing from the juncture of leaf and stem and in the Northeast are produced continually from late June until November. The small, mealy-white seeds grow in dense clusters. Lamb's-quarters is an annual. A moderate-sized plant can and does ripen thousands of seeds, and the germination rate is very high. The plants are not fussy and manage to thrive in the most in-

LAMB'S-QUARTERS:
Two upper branches from different plants,
showing variability of leaf,
both approximately life size.

hospitable places, where they may be the only green living things in sight. Lamb's-quarters is an alien, originally introduced from Europe. It has become naturalized and is so common and widespread in the United States that it grows everywhere and is ubiquitous in the city.

Historical Lore, Legends, and Uses: If you stand on a street corner waiting, perhaps, for a bus and you happen to notice something green flourishing in a crack in the sidewalk, between the lamppost and the cement, you'll most likely be looking at Lamb's-quarters. This plant grows all around us, but most people do not notice it and are unaware of its name or anything else about it. Yet it is a fascinating plant.

The folknames tell a great deal about it. "Lamb's-quarters" refers to the shape of the leaves, which reminded country folk of the plump hindquarters of a lamb. Goosefoot, another common folkname, is a literal translation of the plant's generic Latin name *Chenopodium,* alluding to the resemblance between the triangular-shaped leaves and the webbed foot of a goose. Dirtweed, Dirty Dick, and Midden Myles are amusing (if unappetizing) English folknames that refer to the plant's fondness for growing on manure heaps, where it produces a bumper crop. Pigweed is a Canadian folkname; in that country the plant is grown as a food crop for pigs. They are supposed to be very fond of it, although, curiously, it is mentioned in several old herbals as being fatal to pigs who ate it.

In northern China, Lamb's-quarters was used medicinally to relieve burns and insect bites. In Europe, however, Lamb's-quarters has been known and used as a potherb since very ancient times. "Allgood" was a very old name for it. The early settlers are said to have brought it with them to America and grown it in their gardens. There was even an old English verse that alluded to its tastiness: "Boil Myles in water, chop him in butter, and you'll have a good dish."

The native Americans of the Southwest were aware of the food value of Lamb's-quarters. A botanist traveling in the area during the 1930s observed it growing in every pueblo he visited. The Navajos called it *Quelite* and cultivated it for food. Native American tribes in Arizona and New Mexico prepared a flour from the ground seeds, which formed a highly nutritious staple in their diet.

Suggested Uses: Lamb's-quarters is one of our best wild potherbs. It has a delicious taste, is easy to find, and deserves to be better known and appreciated. It is closely related to beets, chard, and spinach and contains 11,600 IU vitamin A and 80 milligrams vitamin C per 100-gram serving. The recommended daily allowance is 5,000 IU vitamin A and about 75 milligrams vitamin C.

Lamb's-quarters can be gathered practically at your doorstep from early spring to late fall. The tastiest parts are the growing tips and the young leaves. Pick more than twice as much as it appears you need, since they cook down to about half. The leaves are excellent raw in salads.

SAUTÉED LAMB'S-QUARTERS

Wash leaves and add a little water to the pot so they steam-cook rather than boil. Cook about 10 minutes, watching to make sure they do not scorch. Drain off any liquid and save it to add to soup. Melt butter. When hot, add Lamb's-quarters and stir to coat the leaves well. Season with salt and freshly ground pepper. Serve hot.

This is very good as it is, but if you wish you can sauté garlic or chopped onion in the butter before adding the leaves.

LAMB'S-QUARTERS, CUCUMBER, AND TOMATO SALAD

This salad will serve two generously.

1 large ripe tomato, thinly sliced, 1 medium cucumber, thinly sliced, 2 or 3 cups chopped Lamb's-quarters, salt and freshly ground pepper, chopped scallions (optional), oil and vinegar dressing

Arrange a layer of tomatoes on a platter, then a layer of cucumber slices. Top with the chopped Lamb's-quarters, salt and pepper, and scallions, if desired. Pour salad dressing over it.

MEXICAN TEA *(Chenopodium ambrosioides)*

Folknames: Wormseed, Epazote, American Wormseed, Wormseed Goosefoot, Jesuit's Tea, Herba Sancta Maria, Spanish Tea.

Location: Mexican Tea grows everywhere, in parks, roadsides, vacant lots, and waste areas.

Botanical Description: Mexican Tea grows from one to four or more feet tall, but the stems are weak, and tall plants frequently bend down and grow along the ground. The leaves on the lower stems are between one and three inches long, narrow, oval in shape, and pointed at the tips. The leaf edges are toothed. Leaves on the upper stems are much smaller and narrower, and their edges may be toothed or smooth.

Mexican Tea blooms from middle or late July to October or November in the eastern United States. The flowering stalks are arranged at the joint of a leaf and stem, with minute, inconspicuous, yellow-green flowers densely clustered on the stalks. A large plant is capable of producing tens of thousands of seeds; this fertility, along with an ability to tolerate poor soil and harsh environments, explains its abundance. Mexican Tea is generally considered undistinguished looking and not likely to be noticed. Most people in the city have seen it many times without realizing it. A peculiarity of Mexican Tea that does distinguish it from other plants and helps identify it, is the pungent odor of the crushed leaves.

Mexican Tea is native to South America and Mexico. It has become naturalized and grows throughout the United States. It is a plant of warm climates but has adapted itself and thrives as far north as New England. It is a perennial, and reproduces from seeds.

Historical Lore, Legends, and Uses: Most of the folknames for Mexican Tea are South American in origin, and usually refer to its use as a beverage. The word *Epazote,* however, is derived from the Indian words *Nahuatl,* meaning "rank-smelling," and the words *epatl* and *tzotl,* meaning "animal like a skunk." The names allude to the pungent aroma of the fresh plant which, obviously, not everyone found agreeable.

Until fairly recently, Mexican Tea (*Chenopodium ambrosioides,* variety *anthelminticum*) was an important drug plant. It was a powerful vermifuge, quickly and efficiently expelling roundworms and hookworms, two intestinal parasites responsible for a great deal of debilitating illness in humans. A drug known as chenopodium oil was distilled from the seeds, which contain the most concentrated amounts of it, though it is present throughout the plant. This oil does not actually kill the worm, but paralyzes it. A purgative is administered afterward to remove it from the body.

At one time, a great deal of the world supply was grown in the United States, with the best-quality oil being obtained from plantations in Maryland. Chenopodium oil tasted awful and frequently had serious, sometimes fatal, side effects especially in children, and has been replaced in most areas by safer and more effective drugs.

During the nineteenth century, Mexican Tea was considered useful for treating asthmatic conditions and menstrual difficulties. At the present time, in South Carolina, Mexican Tea is still employed as a rural home remedy. A decoction of the whole plant is used to bathe feverish babies.

This plant is still employed medicinally in other parts of the world. The oil from the seeds is used to treat amoebic dysentery in several Asian countries and is applied externally for ulcers and fungus infections of the skin. A tea prepared from the plant is drunk to relieve stomach upsets and asthma attacks, and a decoction of Mexican

MEXICAN TEA:
*Portion of upper stalk with seeds,
approximately life size.*

Tea is used as a wash for inflammations caused by insect bites. The fresh or dried leaves are eaten with salt to relieve menstrual cramps.

Suggested Uses: The leaves of Mexican Tea are used in Mexico, Guatemala, and Peru as a condiment to add flavor to bean, corn, and seafood dishes, and the seeds are used to season meats. A tea prepared from the leaves and drunk after meals is a popular digestive beverage in Mexico and Bolivia. Brazilian peasants add large quantities of the dried plant to mattress stuffings and place sprigs in drawers and chests to repel parasitic insects.

Mexican Tea is a good source of iron, calcium, vitamin A (carotene), riboflavin, and ascorbic acid.

FRIJOLES DE OLLA

In *The Cuisines of Mexico* (Harper & Row, 1972), author Diana Kennedy discusses the culinary aspects of Mexican Tea. She calls it by its Indian name, *Epazote.* It is described as being an acquired taste and best flavored in the fall. *Epazote* should be used fresh rather than dried, and a plant can easily be grown in a pot indoors for winter use. She writes, "There is no substitute for it. It adds a particular distinctive flavor." This is her recipe.

1 pound dried black beans, 10 cups water, 1/2 onion, chopped, 2 tablespoons lard, salt, 2 large sprigs Epazote (Mexican Tea)

Rinse beans and pick over. Cover with water; add onion and lard. Bring to a boil. Simmer slowly 2 hours, or until beans are tender. Do not stir. When tender, add salt to taste, and *Epazote* leaves. Simmer 30 minutes longer. Eat the next day.

MILKWEED *(Asclepias syriaca)*

Folknames: Silkweed, Fluxroot, Butterflyweed, Cottonweed, Wild Cotton, Virginia Silk.

Location: Roadside, waste areas, vacant lots, open places.

Botanical Description: Milkweed is a common and familiar plant in the city. It is very tall, frequently reaching a height of six feet or more, and usually grows in clumps. The large leaves are four to six inches long and oval in shape. The leaf edges are smooth and untoothed. The undersides are downy, with prominent mid-ribs and veins.

In the eastern United States, milkweed blooms during July and August. The flowers are borne at the ends of the stems in large purple-brown clusters and have a rich, sweet fragrance. The individual flowers have an unusual, complex structure best viewed close up with the aid of a magnifying glass.

The plants are perennial, reproducing by seeds or by spreading underground rootstocks. When the ripe pods split open in late summer, the seeds are carried off on fluffy white "parachutes" at the faintest stirrings of a breeze. Stalks and empty pods often persist all winter.

Milkweed is native and widespread throughout the United States. All parts of the plant contain a milky white sap, sticky and bitter to the taste. It flows abundantly when any part of the plant is injured or broken.

Historical Lore, Legends, and Uses: Milkweed was named *Asclepias* in honor of the Greek god of medicine. *Syriaca,* the species name, is an old one and a bit mystifying, since this particular Milkweed did not come from Syria and is indigenous to the United States.

Native Americans were acquainted with the medicinal virtues of the plant and may have been responsible for passing their knowledge on to the early settlers. The Oregon and other western tribes called the plant *Nepesha.* They used the root of this and other species to treat asthma, dropsy, dysentery, and enteritis. The Lenape used Milkweed leaves mixed with tobacco and smoked the mixture to relieve attacks of asthma. The juice was taken internally to cure dysentery and jaundice and applied externally to remove warts.

The Algonquins had a curious use for Milkweed. It was pounded, steeped in water, and made into a tea reputed to be a sure cure for epileptic fits occurring in persons born during certain phases of the moon.

Nineteenth-century American physicians considered it to be an important medicinal plant, particularly efficacious in lung and chest diseases. One doctor observed eloquently: "Milkweed is indeed a valuable remedy, acting safely without stimulating the body, always beneficial in pneumonia and suitable for complaints in children. Its valuable properties, many of which are well-attested, entitle it to general notice to become an article of commerce kept in shops." The powdered root and milky sap were used. A tea prepared from the roots relieved asthma and was given as a mild sedative for nervousness.

Milkweed root is believed strongly emetic, and the mature foliage can cause stomach upsets. In the South, a closely related species, Butterflyweed (*Asclepias*

MILKWEED: (*Below*) *upper stalk with flower buds,*
(*right*) *young immature seed pod,*
and (*far right*) *young shoot, all approximately life size.*

tuberosa), which has orange flowers, was used. This plant was called Fluxweed and Pleurisy Root. Milkweed and Fluxweed are no longer used by physicians today, but the white sap is considered a "sure-fire wart cure" in rural areas.

At one time, Milkweed plants were grown and harvested for their fluffy down. Forty or fifty plants yielded one pound of fiber, which was used to make paper and felt. The staple is too short to spin by itself, but was successfully combined with wool, cotton, flax, and raw silk.

Milkweed and the monarch butterfly have an intimate and fascinating relationship with each other. Indeed, the popular folkname Butterflyweed alludes to the fact that these butterflies are invariably seen fluttering around the flowers.

The tongue of the monarch is especially adapted for reaching the nectar of Milkweed flowers, thus ensuring pollination. The larvae feed exclusively on Milkweed leaves; this diet makes them bitter, and birds will not eat them. The viceroy butterfly has taken good advantage of this. It mimics the colors of the monarch and so closely resembles it as to be indistinguishable. Predators cannot tell one from the other, so the viceroy escapes being eaten.

Milkweed is by far one of the very best edible greens in the city. However, *only* the parts of the plant specified should be eaten. Parboiling (specified in all the recipes) removes all traces of the bitter white sap.

Suggested Uses

YOUNG MILKWEED SHOOTS

Gather them in early spring, starting in May, *when no more than eight inches high.* They somewhat resemble asparagus in appearance but have a special flavor of their own. Trim off the root ends and stem leaves; the leaves at the tip may be left on. Boil the shoots in water to cover for 5 minutes, drain, and cover with freshly boiling water. Boil for a few minutes more or until tender, but they should be crisp. Drain and serve with lots of melted butter.

MILKWEED FLOWER BUDS

These develop as summer advances. Look for them in June and early July. They are tightly clustered at the ends of the stems and resemble miniature broccoli flowerets. Prepare them by removing and discarding all leaves. Drop into boiling water and cook for 5 minutes. Drain, pour boiling water over them, and cook for 5 minutes more. Drain well and serve with melted butter. Cooked Milkweed buds have a delicious, sweet taste and are lovely to look at because they turn a beautiful jade green.

These are edible but take a bit of effort to prepare. They must be harvested toward the end of the summer when the pods are immature, only about three inches long. Even then, the green outside shell may be tough. The undeveloped silk and seeds inside are very good, however, and have a sweet, nutty flavor.

A good way to prepare them is to split the pods open. Discard the green outer covering and poach the insides in water just to cover. When soft, add cream, butter, salt, and freshly ground pepper to taste, and mash together gently. You will have an excellently flavored Milkweed purée. Add a bit of fresh lemon juice if you wish.

In the fall, after the seeds are gone, the Milkweed pods remain on the tall stems. They dry in subtle tones of gray, green, and brown, and twist or curl in a dramatic manner. They are sold (expensively) by florists, often disfigured by spray paint in various (and awful) shades of pink, green, or whatever. Milkweed pods are handsome and decorative just as they are, and make dramatic indoor decorations massed by themselves in a large vase or jar, added to other dried plants in an arrangement, or used individually as natural Christmas tree ornaments.

MOTHERWORT *(Leonuris cardiaca)*

Folknames: Lion's Ear, Lion's Tail, Throw-wort.

Location: Under hedges, in vacant lots and waste areas.

Botanical Description: Motherwort grows from two to four feet tall and has stout, erect, branching stems. These stems are square and are characteristic of many members of the mint family, to which Motherwort belongs. The leaves on the lower stems are shaped like a maple leaf, having three prominent points, a triangular shape, and toothed margins. The leaves on the upper stems are smaller and narrow and usually lack points and teeth. In the Northeast, Motherwort is in bloom during July and August. The flowers are pink, borne in whorls at the juncture of leaves and stem. A perennial, Motherwort reproduces easily from seeds.

Motherwort is an alien. It was brought from Europe, where it was a valued medicinal plant. It is likely that it was grown in gardens. Motherwort eventually escaped cultivation and has become naturalized throughout most of the United States.

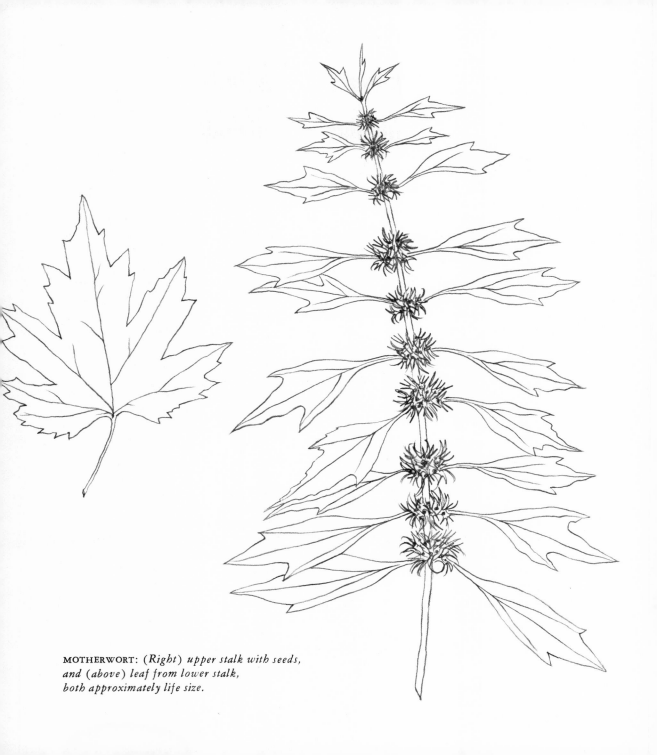

MOTHERWORT: (*Right*) *upper stalk with seeds,
and* (*above*) *leaf from lower stalk,
both approximately life size.*

Historical Lore, Legends, and Uses: The ancient use of Motherwort as the favorite herb for childbirth and heart ailments is reflected in both its scientific and its popular names. The Greek physician Galen recommended it for difficult childbirth and heart disease. The plant was called *Cardiaca* because it was believed valuable against "the infirmities of the heart." William Coles, a seventeenth-century physician and herbalist, believed Motherwort was not known to the ancients, because, as he puts it, "there is some clutter amongst them about it." He was referring to the fact that several classical scholars and naturalists called the plant by a different name, and it appears likely that there was not sufficient information in his time to positively identify the plant these writers called Motherwort. He concludes: "We in English call it Motherwort and not without good reason, for it is of great virtue to help the mother as well as the heart." John Gerard (in his inimitable style) commented that "divers commend Motherwort against the infirmities of the heart. It is judged to be so forcible that it took his name Cardiaca of the effect."

Cardiaca has been retained as the species designation for Motherwort, but the derivation of the generic name *Leonuris* is a bit more obscure. *Leonuris* is derived from the Greek and means "lion's tail." Some believe this alludes to a fanciful resemblance between the plant and the tail of a lion, but Motherwort was a plant of the sun, and the lion is the ancient animal symbol for the sun. It is possible, therefore, that the name *Leonuris* comes from this association.

So greatly esteemed was Motherwort as a women's herb that the plant has been known by this name in practically all the languages of Europe. "There is hardly a more effectual herb for the womb than Motherwort," Dr. Coles observed. "It is most wonderfully used in childbirth, and therefore it has not got its name for nothing." He noted with compassion that a woman in labor was "most commonly incapable of being placed in or on any decoction, or taking anything inwardly," and he recommended applying little bags of the herb, warmed, to the lower abdomen. "There is scarcely a better remedy," he concludes.

For menstrual difficulties, the plant was chopped and a large quantity of boiling water was poured over it. The woman was instructed to sit over this bath in order to absorb the beneficial steam. The powdered herb was taken in wine to expedite labor and the passing of the afterbirth. In addition, it was thought excellent to relieve painful menstrual cramps, convulsions, and the "trembling of the heart." Taken in wine, Motherwort would "drive melancholy vapors from the heart and make one merry, cheerful and blithe."

In the United States the Algonquins may have learned about the virtues of Motherwort from the Europeans, because they, too, used it to treat female disorders. A tea was prepared from the leaves and taken internally.

Motherwort is still listed in the *British Herbal Pharmacopoeia,* and is recommended for the relief of cardiac palpitations associated with anxiety. Combined with horehound, it is used to dispel false labor pains. The plant is very bitter, and because an infusion of the leaves is most unpalatable, modern herbal doctors prepare a conserve of the young fresh tops with honey or sugar.

In Japan, Motherwort was reputed to have the ability to confer long life. It is said that long ago, a village was situated on a hill covered with the plants. A pure stream flowed down through them to the foot of the hill, and the villagers drank from it often. Many of them lived to be more than one hundred years of age, and this longevity was attributed to the Motherwort, whose virtues were absorbed by the water.

It became a custom to drink a beverage prepared from the flowers of Motherwort. A special festival in honor of the plant was held on the ninth day of the ninth month, and the month itself was called *Kikousouki,* "the Month of Motherwort Flowers." The nobility drank a beverage called *zakki* (a forerunner perhaps of the modern sake), and in the countryside the peasants tied flowering sprigs of Motherwort to pitchers of beer, to signify that they wished each other long life.

Modern herbalists consider Motherwort excellent for easing menopausal difficulties.

Suggested Use: According to the early writers, there was no better herb than Motherwort for "strengthening and gladdening the heart." It was usually taken in the form of a conserve or syrup, because of its bitter taste.

MOTHERWORT SYRUP

Pour 2 cups of boiling water over ½ cup fresh Motherwort leaves and young tops. Let steep 10 minutes. Strain. Add 1 cup sugar to liquid, heat slowly, and simmer until sugar is dissolved. Store syrup in the refrigerator. To prepare a drink to "calm the spirits," add 1 or 2 tablespoons of Motherwort syrup to 1 cup of boiling water. This drink is good taken just before retiring.

MUGWORT *(Artemisia vulgaris)*

Folknames: Felon Herb, St. John's Plant, Sailor's Tobacco, Chrysanthemum Weed.

Location: Roadsides, waste areas, vacant lots, grassy strips.

Botanical Description: Mugwort is usually one to four feet tall, but it may grow as much as six feet high in a favorable location. The stems are erect, somewhat woody near the base, and branched. The much-divided leaves are dark green above and silvery and somewhat downy on the underside. In the eastern United States spikes of tightly clustered, minute silver-green flowers begin appearing in late July and continue into September. The foliage is aromatic.

Mugwort is one of the most commonly seen plants in cities. An observer who takes the trouble soon becomes aware that innumerable roadsides, vacant lots, and similar areas are wholly occupied by luxurious stands of this rapidly spreading perennial. The plants are hardy, tenacious, and highly adaptable, able to survive in environments that appear incapable of supporting anything green.

Frequent mowing or cutting of Mugwort plants reduces their height but does not kill them, nor will uprooting the plants eradicate them. Mugwort grows from runners that form dense underground mats, and unless every bit of root is taken up, new shoots quickly grow from any pieces remaining in the earth.

Left to its own devices, however, Mugwort performs an invaluable service. During the first warm weeks of summer, large clumps speedily transform empty lots, waste areas, and similarly bleak places with a living screen; the dreary rubble and trash is mercifully hidden from view in its green and silver foliage.

Mugwort was introduced from Europe to the New World and has become naturalized throughout most of the United States, especially the eastern half.

Historical Lore, Legends, and Uses: The old herbalists claimed that Mugwort was excellent to "comfort the braine," and I believe this to be true. On a hot summer day, when the noxious fumes and stagnant air in the city seem even more oppressive than usual, it really is a "comfort" to crush a few leaves of the plant in one's hand and inhale the clean, pungent aroma. Mugwort lives up to its ancient reputation and certainly has the power to revive the spirits and refresh the senses.

The name Mugwort is derived from the use of the plant in the preparation of beer and ale. Mugwort leaves were added to the kegs for flavor and to prevent souring. Even after the introduction of hops, which quickly supplanted the native herbs long favored by the people, Mugwort continued to be an ingredient in beer.

Mugwort was considered an important medicinal herb, and since ancient times was believed to have a special affinity with women. *Mater Herbarum,* or "Mothers' Herb," was the Latin name for it. According to the Roman naturalist Pliny, the plant was named *Artemisia* in honor of Artemisia, Queen of Caria. This noble lady learned the use of Mugwort for treating women's ailments and taught its virtues to others. An Anglo-Saxon leechbook gives another explanation for the name, old before they set it down in the tenth century: "Verily, of these worts which we name Artemisia, it is said that Charon the Centaur, who first from these worts set forth a leechdom, named them for Diana." Diana (also named Artemis) was goddess of the moon. The moon and this goddess were anciently associated with women and their internal rhythms of menstruation, conception, and childbirth.

The medicinal properties of Mugwort were thought to reside in the fresh leaves and young growing tips. Boiling water was poured over a quantity of chopped plant material, and the ailing woman sat or squatted over the beneficial steam. Absorbed thus into the body, Mugwort was said to ease labor, relieve painful menstruation, and help to heal a variety of uterine disorders. Poultices of the leaves laid on the belly did the same, as did a syrup of the leaves taken in water or wine. A tea was good to strengthen the nerves and quieted those taken with "hysterical fits."

The fresh juice of Mugwort, pressed from the leaves and tops, is frequently listed in the old herbals as an effective antidote for overdoses of opium. More amusingly, it is often mentioned that the steam from Mugwort "makes small children joyful." We are not told if it had the same effect on adults.

In the United States during the nineteenth century, Mugwort was an official drug plant. Physicians used infusions of the fresh plant or the dried, powdered herb. Taken internally, Mugwort promoted menstruation, relieved cramps, and was considered an excellent sedative, laxative, and diuretic.

Today, modern homeopathic doctors prescribe a tea made from the dried leaves of Mugwort as a mild sedative of benefit to menopausal women. This tea also increases stomach secretions and is supposed to aid digestion.

Mugwort, in a preparation known as *moxa,* has been used by the Chinese for centuries, as both a cure for and a preventer of a wide variety of diseases. Valued by young and old, rich as well as poor, the treatment known as *moxa burning* is by no means extinct and recently has even begun to attract adherents in the West.

To make *moxa,* Mugwort leaves (related varieties such as *Artemisia sinensis* are used as well) are collected in early summer and carefully dried. Then they are crushed, either with a mortar and pestle or by rubbing between the hands, until the woody fibers are separated out and only a fine, cottonlike fiber remains. This material is formed into small cones or cylinders and is then ready for use.

Moxa is placed, often several cones at a time, on specific areas of the body according to points on a special *moxa* chart. It is then set alight with a joss stick or taper and allowed to burn out. This causes a small blister to be raised on the skin, and the *moxa* ash is rubbed into it.

At one time, *moxa* was burned to ensure that a newborn infant would survive babyhood. It was burned on the face between the brows, at the base of the nose, and on the cheeks just below the eyes, when the child was three days old.

Folklore and superstition are bound up with the Mugwort plant to an extent hardly matched by any other herb. No matter how disparate the language or culture, legends similar to one another are encountered in all parts of the world where the plant grows. In most cases, the origin of the belief can no longer be traced. No matter how bizarre, these legends are frequently accepted as the truth and are perpetuated by otherwise accurate and discriminating observers.

Perhaps the oldest superstition associated with Mugwort is its magical efficacy in warding off exhaustion and protecting against evil. Pliny said that travelers who carried Mugwort would never become weary, nor would "evil medicine" (probably witchcraft) harm them. This naturalist died in A.D. 79; fifteen centuries later, credence was still given this statement.

The Anglo-Saxons used Mugwort medicinally; with them it seemed to be a panacea for all the ills flesh was heir to, and they valued its magical properties as well: "It puts to flight devil sickness and in the house in which he, the man of the house hath it within, it forbids any evil leechcrafts and turns away the evil eyes of evil men." The following poem honoring Mugwort was written in a tenth-century leechbook:

> *Have a mind, Mugwort*
> *What thou mentionedst*
> *What thou preparedst*
> *At the prime telling*
> *Una thou hightest [Una commanded you]*
> *Eldest of worts,*
> *Thou hast might for three*
> *And against thirty;*
> *For venom availest*
> *For flying vile things [probably plague]*
> *Mighty 'gainst loathed ones*
> *That through the land rove.*

MUGWORT:
*Upper part of leafy stem,
slightly larger than life size.*

Separated as we are from this poem by almost ten centuries, some of the references are a bit arcane. But it seems clear from the verses that the Anglo-Saxons believed Mugwort was powerfully effective against diseases of natural or supernatural origin, and in the poem, this well-beloved "Eldest of worts" is entreated to not forget its promise to aid humankind.

Mugwort was renowned as an herb of life probably because of its hardiness. In Italy a form of divination was practiced with it: a green sprig of Mugwort was slipped under the pillow of someone gravely ill. If she (or he) fell asleep, she would recover; but if not, she would die.

In Scotland, the life-preserving attributes of Mugwort were revealed to the people by a mermaid who, as she watched the funeral procession of a young girl who had died of consumption, exclaimed:

> *If they eat Nettles in March*
> *and drink Mugwort in May,*
> *so many fine maidens*
> *would not go to the clay.*

In medieval times Mugwort, formerly an herb of Diana, was placed under the protection of St. John. If it was gathered on St. John's Eve and worn in a garland, Mugwort would be even more than usually powerful to avert disease, bad luck, demonic possession, and the evil eye.

In Sicily, crosses were fashioned of Mugwort plants and placed on the roof of the house on Ascension Eve for the Lord Jesus to bless as He reascended to heaven. The crosses protected the home for a year and were afterward placed in the stable, where their presence helped make unmanageable animals docile.

In China, Mugwort plants were believed to have the power to keep noxious influences and evil spirits at bay. During the Dragon Festival (which took place on the ninth day of the ninth month) Mugwort was hung in the most important room of the house. There is an even older Chinese custom of hanging Mugwort outside, on each doorpost. According to legend, the famous rebel Huang Chao had such reverence for the plant that he ordered his soldiers to spare any family that hung Mugwort thus.

This same Mugwort, so greatly valued as a protector against evil spells, was a frequently used and supposedly potent ingredient in these same spells and enchantments it was supposed to offer protection from. Many old *grimoires* and similar books of magic have recipes for using it.

The Pennsylvania Dutch knew Mugwort as *Aldy Fraw,* "Old Woman." They planted it in their gardens but were careful never to put it near another variety of

Artemisia they called *Alder Mon,* "Old Man." It was said that the two plants did not get along at all, and that "Der ald mon macht die ald fraw doet!" That is, "The Old Man will kill the Old Woman!"

There is a Russian folktale about how Mugwort, called *Zabytko,* got its name: one day, a young girl went into the forest to gather mushrooms. She stumbled and fell into a deep pit, which turned out to be the abode of serpents. They did her no harm and actually took care of her, even showing her a certain stone that glowed in the dark and magically provided the snakes (and the girl herself) with nourishment.

She stayed with them throughout the winter, and when spring came, the creatures formed a ladder with their intertwined bodies, and the girl was able to climb out of the pit. As a parting gift, the Queen of the Serpents gave her the power to understand the language of the plants but warned her never to name Mugwort, for if she did, her magic gift would immediately be lost to her forever. One day long afterward, as she was walking with her lover, he asked suddenly, "What is the name of the plant that grows in the fields, beside the little footpaths?" Taken by surprise, she answered, "Mugwort." From that time on the maiden forgot the speech of the plants, and the Russians have called Mugwort *Zabytko,* the herb of forgetfulness.

My own favorite Mugwort legend is of ancient Germanic origin: it was long believed that if Mugwort were dug up at noon on St. John's (Midsummer) Day, a burning coal would be found under the root. Carried away in secret, this magic coal would be a certain remedy against evil and might eventually even turn into gold.

This legend is reminiscent of the many "impossible tasks" that abound in folklore. Mugwort grows so abundantly that finding the particular plant with the magic coal at its root, like attempting to sweep away the sands of the beach, would have taken an infinity of time and effort. The legend persisted and was already several hundred years old when John Parkinson, the renowned seventeenth-century herbalist, decided he'd had enough: "Many idle superstitions and irreligious relations are set down, both by the ancient and later writers, concerning this and other plants, which to relate were both unseemly for me and unprofitable for you. I will only declare unto you the idle conceit of some of our later days concerning this plant." Dr. Parkinson proceeds to describe an experience related to him by someone whose opinion he had once respected. This individual turned out to be someone "who glorieth to be an eyewitness of this foppery, that upon St. John's Eve there are coals to be found at midday under the roots of Mugwort which after or before that time are little or none at all, and are used as an amulet to hang about the neck of those that have the falling sickness, to cure them thereof. But oh," he declares mournfully, "the weakness and frail nature of man! Which I cannot but lament, that is more prone to believe and rely upon such impostures than the aid of God!"

Dr. Parkinson would have been pleased to hear, some forty years after his declaration, a terse pronouncement by a naturalist named Paul Babette: "These coals," he said, "are old dead roots." That was the end of the matter.

Suggested Uses: Mugwort is closely related to Wormwood, Rue, and Tarragon and has been used as a culinary herb in Eastern Europe. Dried and crumbled Mugwort leaves were added to the stuffing for geese and other rich, fatty fowl to add flavor and aid digestion.

POTATO DRESSING FLAVORED WITH MUGWORT

This recipe makes enough to stuff a 4- to 6-pound fowl. Ingredients may be increased accordingly for larger birds. This dressing is excellent with fat roasting chickens, capons, ducks, geese, and game birds.

3 all-purpose medium potatoes, 1 large onion, chopped,
3 tablespoons chicken fat or vegetable oil, 2 tablespoons finely chopped fresh parsley,
dried Mugwort leaves, salt and freshly ground pepper

Boil potatoes in their skins until cooked, then peel and dice them and keep them warm. Fry onion in fat until golden brown. Mix onion, fat, and potatoes together. Add parsley and dried, crumbled Mugwort leaves to taste. Mugwort is strong, so begin with about 1 teaspoon and add more if you wish. Season to taste with salt and pepper.

To dry Mugwort leaves: I prefer to dry the leafy tips of the Mugwort rather than the leaves on the lower stems, because their flavor is better. Place the leaves one layer deep on trays in a room with good air circulation, away from direct sunlight. When they are thoroughly crisp and dry, store in jars with tight-fitting lids, away from the sun.

TEA LOAF FLAVORED WITH MUGWORT

Mugwort, fresh or dried, is the special herbal seasoning for this quick bread. It is best served warm or toasted, with butter.

2 cups unbleached flour, 1½ teaspoons baking powder, 1 teaspoon salt, ¾ cup water,
1 egg, beaten, 2 tablespoons salad oil, 1 tablespoon finely chopped dried Mugwort or
2 tablespoons finely chopped fresh Mugwort, ½ cup grated Cheddar cheese,
1 tablespoon caraway seeds (optional)

Preheat oven to 350°. Sift flour, baking powder, and salt together into a large bowl. Mix water, egg, and salad oil together and pour into dry ingredients along with chopped Mugwort leaves (see above for how to dry Mugwort), cheese, and caraway seeds, if they are used. Do not stir until all ingredients are added. Stir only enough to moisten; *do not overmix.* Spoon batter into buttered 9½-by-5-by-3-inch loaf pan. Bake for about 45 minutes, or until a cake tester inserted in the middle comes out clean. Cool in pan for 5 minutes, then remove from pan and cool on a rack. For muffins, spoon batter into buttered muffin tins and bake about 25 minutes. Makes approximately 12 muffins.

A few handfuls of fresh Mugwort leaves can be added to the bath to relieve tired legs and feet.

MULLEIN *(Verbascum thapsus)*

Folknames: Clown's Lungwort, Bullock's Lungwort, Our Lady's Flannel, Adam's Flannel, Indian Flannel, Feltwort, Velvet Dock, Velvet Plant, Woolen, Blanket Herb, Beggar's Blanket, Candlewick Plant, Torches, High Taper, Hag Taper, Witches' Taper, Aaron's Rod, Jacob's Staff, St. Peter's Staff, Quaker Rouge.

Location: Roadsides, embankments, vacant lots, and similar open, sunny areas.

Botanical Description: Probably everyone has seen Mullein. It is a pretty plant, particularly conspicuous along highways in the city. Mullein is quite tall. It generally grows to about four feet high, but plants of six feet and more are not uncommon.

Mullein is a biennial and in the first year produces a rosette of large, gray-green, feltlike leaves. The following year a tall, rigid stalk grows from the center of this rosette. This stalk (occasionally branched near the top) is clasped along its entire length by smaller leaves which actually merge with the stalk at their bases. The upper part of this stalk becomes the flower spike. As summer advances, it becomes covered with densely packed buds. In the Northeast, delicate yellow flowers open at random along this stalk from late June until September.

The entire plant is covered with fine, downy hairs that give Mullein its characteristic soft, velvety texture and appearance, and help the plant retain moisture. In the fall the woody stalks are covered with a beautiful starlike pattern of empty seed pods, which turn brown and persist on the stems throughout the winter.

Mullein is an Old World plant. It was introduced from Europe, where it was for centuries an esteemed medicinal herb. The colonists planted it in their gardens; it has since escaped and become naturalized throughout the United States.

Historical Lore, Legends, and Uses: The great respect and love formerly accorded to Mullein can be inferred from the number and variety of the folknames for it, of which the above is but a partial list. Some allude to the feltlike, woolly texture and size of the leaves; others reflect usage.

Mullein was known in Greek as *Flego* and *Fluma,* that is, "to set on fire." According to one writer, "it served as a wick to put into lamps to burn." The leaves were rolled and dried and used as wicks for oil lamps and candles, and made excellent tinder. The Latin names for Mullein were *Candelaria* and *Candela Regia.* "The elder age," observed John Parkinson, a seventeenth-century herbalist, "used the stalks dipped in suet whether to burn at funerals or otherwise, and so likewise the English name *High Taper,* used in the same manner as a taper or torch."

Mullein has been a valued medicinal herb since antiquity. The Greek physician-herbalist Dioscorides was one of the first to recommend its use in curing diseases of the lungs, and it remained thus employed for more than 1,800 years. Infusions of the leaves and flowers were used, and similar preparations were administered to cure lung ailments in swine and cattle.

Mullein was used to treat a variety of other ailments. A wash prepared from the leaves, flowers, and roots soothed sprains, reduced inflammations, and healed wounds. The flowers infused in oil were used to cure hemorrhoids and as a specific cure for earache.

In our own country, several native American tribes used Mullein to cure chest diseases. Since the plant was not native to America, this usage was probably received by them (no doubt along with the lung ailments it was said to cure) from the early settlers. The Navajos called Mullein "big tobacco." They mixed it with regular tobacco and smoked the combination to relieve coughing spasms. It was also believed that this remedy would cure simple mental diseases, the use of evil language, and the thinking of evil thoughts.

The Delaware made poultices by boiling the leaves of Mullein and putting them into cloth bags. Applied to the joints, these poultices were said to reduce swelling and ease rheumatic pains. The Catawba called Mullein "gray leaf." They made poultices of the mashed leaves and used them for treating sprains, swelling, and wounds. For lung and bronchial troubles, the Catawba gathered Mullein leaves from plants that had not yet blossomed and mixed them into a syrup with another plant called "plum root." The Mohegans of Connecticut made a cough medicine by steeping Mullein leaves with molasses.

Mullein was known to the Pennsylvania Dutch as *Wolla Graut.* The Amish eschewed the use of tobacco, but permitted Mullein leaves to be smoked for the relief of asthma attacks. To soothe nasal congestion and sore throat, boiling water was

MULLEIN:
Upper portion of flowering stalk.

poured over the fresh leaves and flowers and the steam was inhaled.

In the United States during the 1800s and early 1900s, Mullein was frequently prescribed for pulmonary diseases and physicians recommended its use for curing diarrhea and severe headache as well. Until quite recently, Mullein was a popular folk remedy in the Ozarks. Mullein-flower tea was drunk to relieve all kinds of chest disorders, from mild colds to pneumonia, and poultices of the leaves soaked in hot vinegar were applied to swellings and sprains.

This last remedy was considered helpful for most painful conditions; an observer who lived among these people for many years remembers being told that a poultice of Mullein leaves was used to ease the pain of wounds caused by bird shot. It also loosened up the pellets and made them easier and less painful to remove.

Mullein was listed as an emollient and demulcent in the 1917 edition of *Potter's Therapeutics,* which noted: "It has long been a popular Irish remedy in pulmonary affections" (asthma and whooping cough particularly). An infusion of the dried leaves in milk was recommended as a valuable expectorant that also eased coughing and improved the general condition. It was useful for cystitis, irritable bladder, and diarrhea as well.

In rural South Carolina, Mullein-leaf tea is still employed (with the occasional admixture of basil and pine needles) to relieve colds and reduce fevers. Applied externally, this tea heals sprains, and a poultice of the leaves mixed with fat is used to bring boils to a head. At one time, large quantities of Mullein were dried and placed in barns to keep mice away from the grain stored there.

The ancient beneficial properties of Mullein have not been entirely discarded. The plant is included in many over-the-counter asthma remedies, and modern homeopathic doctors still prescribe it for treating chest complaints. In Europe an extract of Mullein is used in the preparation of an Old World liqueur called *Altvater Jaegnerdorf.*

It ought hardly to surprise anyone that Mullein, companion to humankind for more than two thousand years, became endowed with various occult and supernatural powers. Not only is this the case, but these properties long predated the medicinal use of Mullein.

The Anglo-Saxons were quoting an ancient legend when they observed: "Of Feldwort [as they called Mullein] it is said that Mercurius gave this wort to Ulixes the Chieftain when he came to Circe, and after that he dreaded none of her witchcraft." They add, "If one weareth with him one twig of this wort, he will not be terrified with any awe, nor will a wild beast hurt him, or any evil come near."

Much later, it became the custom in parts of Europe to dip Mullein stalks in tallow and burn them to frighten away witches. The names of saints may have been invoked for additional assistance in warding off the evil host, and this may account for names such as St. Peter's Staff and Aaron's Rod. (Aaron, it will be remembered, is described in the Book of Exodus as using his magic staff to overcome the sorcerers of Pharaoh.)

With the ambivalence so frequently encountered in superstitious beliefs, Mullein had a reputation for being a favorite plant of witches. In England, Mullein torches were burned to illuminate their Sabbat revels, alluded to by folknames such as Hag Taper and Witches' Taper. Mullein was a plant of Saturn, the planet of evil, who ruled all poisonous herbs and plants of ill repute that were employed by the followers of the devil.

Mullein leaves were worn as charms to ensure conception. John Gerard apparently believed they could have the opposite effect as well, for he advised that Mullein worn in the shoe "brings down in maidens their desired sickness" (which sounds like a euphemism for an abortifacient).

As might be expected, Mullein was used in love divination. A girl sought out a Mullein plant and named it for her lover. She then bent the stalk toward her home and visited it from time to time to observe how it grew. If it remained bent toward her house, he was faithful; if not, he was untrue. As a cosmetic herb, Mullein was used as far back as Roman times. A wash reputed to restore to its original color hair that had turned gray was prepared from ashes of the plant. During the Middle Ages a rinse made with the flowers was said to keep the hair blond, but it had to be used for a long time. Young girls rubbed their cheeks with a leaf to make them pink; much later, Quaker women, forbidden the use of rouge, did the same.

Suggested Uses: The dry seed stalks of Mullein gathered in the fall are among the most decorative wild plants for indoor decoration. They can be combined with other dried plants, but are even more effective when massed by themselves in a large jug. This arrangement looks especially handsome next to a fireplace. Individual stalks of Mullein can even be used as tapers to start the fire.

An unusual recommendation involving Mullein is mentioned in an old volume of the *Standard Cyclopedia of Horticulture* (1933 edition). The first-year rosette, it suggests, can be potted as a houseplant, for the sake of its attractive shape, color, and texture.

MULLEIN-FLOWER TEA

Pour 1 cup of boiling water over 1 heaping tablespoon of Mullein flowers. Let steep 10 minutes, strain, and sweeten with honey if desired.

This is a pleasant, nutritious drink that, taken at bedtime, soothes irritated bronchial passages and relieves coughing. Modern herbalists also recommend this drink to relieve diarrhea in adults.

Combine 2 tablespoons of Mullein flowers (or you may substitute chopped fresh Mullein leaves if you wish) with 1 pint of milk. Heat to the scalding point and let stand until warm. Strain and sweeten with honey.

Note: Mullein drinks should be strained through coffee filter paper, to remove the fine hairs that cover the entire plant. These are irritating to the mouth and throat.

MUSTARDS *(Brassica nigra, B. alba)*

Folknames: Black Mustard, White Mustard, Scurvy, Kerlock, Warlock.

Location: Roadsides, vacant lots, waste areas.

Botanical Description: Black Mustard (*Brassica nigra*) is an erect, branched plant, generally from one to three feet high. The stem is usually bristly near the base, smooth above. Leaves on the lower stems are egg-shaped and bristly; those on the upper stems are smooth, narrow, and much smaller. Yellow flowers are borne at the end of the stems and the juncture of leaves and stems from late June through August in the northeastern United States.

The seed pods of Black Mustard resemble miniature pea pods. They are smooth, erect, have tiny "beaks" at their tips, and hug the stalk. When ripe, the seeds themselves are dark red-brown to black, hence their name. Seed pods and flowers are frequently present on the stalks at the same time.

White Mustard (*Brassica alba*) is similar in appearance to Black. The yellow flowers are larger and the leaves are more deeply lobed. The seed pod is more round, bristly, and has a longer, flat, sword-shaped "beak." The seeds are light-colored and larger than Black Mustard.

The mustards described above are two of several members of the Mustard family that are often found in urban areas. The crosslike or cruciform arrangement of the characteristic four-petaled flower has given the scientific name *Cruciferae* to this family. Other members include cabbage, radish, and turnip.

Black and White Mustard are both alien and were introduced from Europe. They have become naturalized and grow throughout the United States and are especially abundant in the Northeast and California. They are annuals and readily reproduce from seeds. The seeds of most species are capable of remaining viable for at least five years. Charlock or Wild Mustard (*Brassica kaber*) has seeds that can remain dormant for fifty years or more.

Historical Lore, Legends, and Uses: Mustard has a truly ancient history as a medicinal plant and a condiment. A Sumerian cuneiform tablet dating from approximately 2100 B.C. lists a plant that has been identified as White Mustard. (This kind of documentation is extremely rare and brings great joy to the hearts of herbalists.) The tablet contains a description of the part of the plant to be used medicinally, as well as instructions for preparing and administering the medicine.

It is not known when Mustard was first employed as a condiment, but the Romans have been credited with blending the crushed seeds with must to make a sauce. "Mustard" is an old name, believed to have derived from the Latin *mustem,* meaning "must" (fermenting grape juice that has not yet become wine), and *ardens,* which means "burning" and alludes to the pungent hotness that pervades the entire plant but is especially concentrated in the seeds.

For centuries Mustard was believed to have the power to cure almost all diseases flesh was heir to and was used to treat everything from sore feet to bubonic plague. The Romans introduced Mustard into Britain, and by the tenth century it had become an important medicinal plant. The tenth-century Anglo-Saxon *Leechbook of Bald* prescribes two headache remedies containing Mustard. In the first, a vesselful of green rue leaves and a spoonful of Mustard seeds were rubbed together and the white of an egg was added, "that the salve may be thick." This was daubed with a feather "on the side which is not sore." To relieve "half a head's ache," laurel and Mustard were mixed together, vinegar was added, and this mixture was to be smeared on the *sore side!* A leechdom (remedy) for bronchial congestion was compounded with Mustard, water, vinegar, and honey. "Boil them together cleverly and strain. Cool, then give it to him frequently to swill his jowl, that he may comfortably break out the ill flegm [*sic*]." This particular remedy probably worked quite well. It is certainly more sensible (if less fanciful) than the headache cures!

Black Mustard contains two chemical compounds, myrosin and sinigrin, that react with each other in the presence of water to form volatile oil of Mustard. In its pure state, this oil is one of the most powerful caustic agents known. A drop applied to the skin will cause instant and severe blistering. Mustard oil severely damages capillary walls, causing "Mustard burns" that heal very slowly.

When carefully controlled, however, this oil (and other preparations of Mustard as well) had important medicinal uses. Volatile oil of Mustard was formerly an ingredient in a drug that, taken internally, was reported highly successful in promoting the absorption of scar tissue. Compound liniment of Mustard, a blend of Mustard oil, alcohol, and other ingredients, has long been a favorite with athletes for the relief of strained, aching muscles. Mustard is a powerful germicide, and surgeons used to disinfect their hands with a paste of Mustard seeds and water.

Last but not least is the well-known Mustard plaster. This homely remedy for pulmonary congestion, first described and recommended by the Greek physician Dioscorides almost two thousand years ago, is still employed. Mustard plasters came to be considered of particular benefit and value in the treatment of pulmonary ailments, with Black Mustard being the preferred species for this purpose.

When I was a child and developed bronchitis, my mother usually applied a Mustard plaster. Powdered Mustard was mixed with flour (otherwise it was too strong), water was added, and this paste was spread between two pieces of soft, old cotton sheet, wrapped in flannel, and laid on my chest. I remember the peculiar smell (which I disliked) and the care with which the Mustard plaster was checked: it had to be kept in place long enough to impart a dose of therapeutic warmth to the chest—this was indicated by the degree of redness—but had to be removed before it irritated the skin. Mustard plaster if improperly applied was quite capable of producing blisters on the skin, several layers of fabric notwithstanding. But one did feel better and breathe easier afterward.

At the present time, Black Mustard is listed in the *National Formulary* of the United States and can be purchased in powdered form from any drugstore. In addition to its proven usefulness as a counterirritant in easing the pain of various chest and lung congestions, Mustard powder taken in water relieves indigestion and is an effective emetic for certain kinds of poisons that have been ingested.

The Romans carried Mustard sauce, their beloved condiment, with them to Britain, and by the sixteenth century Mustard was a popular seasoning in England. Shakespeare mentions it frequently. In his day, dry, powdered Mustard was shaken onto meats with a Mustard shaker, and in parts of England it is so used today.

By the seventeenth century, the crushed seeds were mixed with vinegar and praised as "an excellent sauce for meat and fish." One writer commented helpfully:

If there be anyone who does not think this ordinary Mustard delicate enough, or less pleasing to the palate or stomach, let them take two ounces of Mustard, half-an-ounce of cinnamon, crushed, and make it into little balls or cakes with honey and vinegar and dried

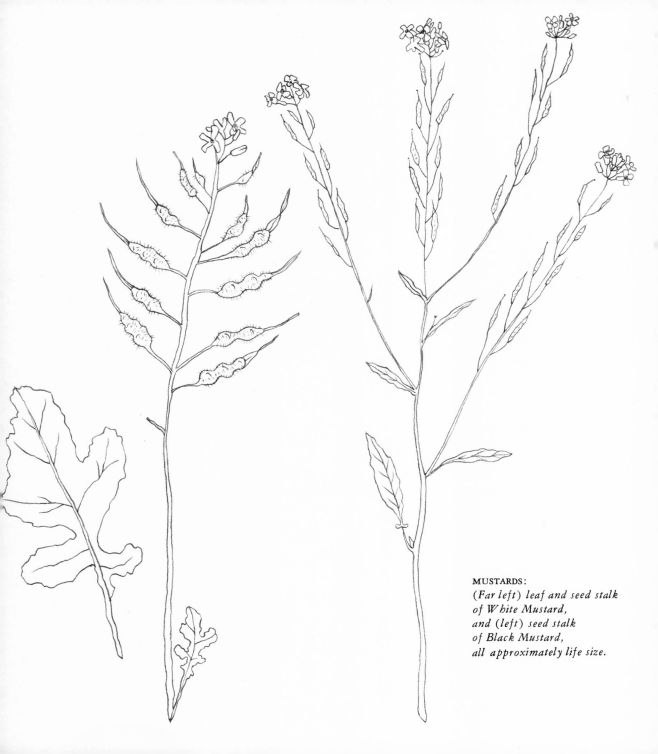

MUSTARDS:
(Far left) leaf and seed stalk
of White Mustard,
and (left) seed stalk
of Black Mustard,
all approximately life size.

in the sun. It will keep a long while until use shall be made of it; by relenting it with vinegar, of which so simple a sauce has not its parallel.

This particular recipe may have originated in Tewksbury, where Mustard was cultivated. Tewksbury Mustard was accounted the best in England, and great quantities of it were brought to London for resale elsewhere.

Mustard balls remained the standard form in which the condiment was prepared until Mrs. Clemens of Durham invented a particular method of grinding the seeds. "Mustard flour," as she called it, was presented to King George I by the lady herself, in the form of a sauce mixed with wheat flour and water. So delighted was the King with this new preparation that he granted her a royal patent. Durham Mustard became famous, and Mrs. Clemens died a wealthy woman.

In addition to the useful seed, Mustard greens were considered a fine potherb. In his delightful volume, *A Discourse on Sallets,* John Evelyn spoke appreciatively of Mustard. "It is exceedingly hot, not only the seed," he noted, "but the leaf also, of incomparable effect to quicken and revive the spirits, strengthen the memory and expel heaviness. It is a most necessary ingredient to all cold and raw salleting, that it is very rarely, if at all, left out."

John Evelyn was probably describing White Mustard, cultivated in his time (1699) as a salad herb. It was eaten when very young and tender. Mustard greens were valued raw in "sallets" for their antiscorbutic qualities. The greens, like the seeds, were believed to strengthen the stomach, promote appetite, and aid digestion.

Mustard had the curious reputation of especially benefiting singers. This idea persisted for centuries, and most of the old herbalists mention it. "Crush the seeds and mix them with honey," advises one writer. "Make into little balls, swallow one every morning, and in a short time one will have a clear voice to sing with." Another herbalist recommends this preparation to "those addicted to singing." It would, he said, "clear their voices very much in a short time."

Fabulous virtues were attributed to Mustard. The leaves were "good to be laid on the heads of them that have the drowsy evil, or forgetfulness called Lethargy." The juice of fresh Mustard plants rubbed on the temples, forehead, and nostrils "warmed and quickened those dull spirits that are as it were asleep or almost dead, for by its fierce sharpness it proceeds to the brain and purges it by sneezing."

Chewing Mustard seeds "cleansed the brain," and eating them sharpened wit and memory, though for this last purpose some held it more efficacious to rub the soles of the feet with Mustard seeds ground up in vinegar.

William Coles seemed to have discovered a quality in Mustard that was apparently overlooked by other herbalists: "Mustard seed," he observed, "by itself or in drink, mightily stirs up bodily lust by the heat it causes, being a great help to remove that deadness and stupefaction that possesses the seed and members of generation in those who are of cold constitutions."

The Parable of the Mustard Seed, found in the New Testament and attributed to Jesus, is familiar to many. Indeed, a single Mustard seed embedded in clear plastic is still worn as a charm. Less well known, however, is a touching Chinese parable concerning Mustard seeds; it is attributed to Buddha.

A young mother, inconsolable over the death of her child, visited a holy man and begged him to restore the infant to life. He would do so, he answered, if she brought him a handful of Mustard seeds from a dwelling where no one had died. She agreed to do so, and went on her way. The woman stopped at one house after another, asking each time if there had been any deaths within. "Of a certainty," was the invariable reply. "For the living are few and the dead are many."

Finally, after much wandering, the mother perceived the futility of her quest and the selfishness of her grief. She saw that others had sorrowed over the deaths of loved ones, and she realized that no matter how great the anguish, the dead cannot be restored to the living. The woman returned to the holy man and knelt at his feet. "I did not find the Mustard seeds," she told him, "but I have found your meaning."

Prepared Mustard (of which Black Mustard seeds are the major source) has become one of America's favorite condiments. Where would our national hot dog, beloved of sports fans and children everywhere, be without it? Some enjoy Mustard sweet and mild; others relish a sauce so pungent that it brings tears to the eyes. Domestic manufacturers have their own blends which they usually keep very secret.

In addition to seeds grown in the Plains area of the United States and Canada, Mustard seeds are imported from Italy, Ethiopia, England, Denmark, and the Orient. Imported prepared Mustards from England, France, Germany, and Sweden, to name just a few countries, are stocked by gourmet stores throughout the country.

Suggested Uses

MUSTARD, PREPARED IN THE SEVENTEENTH-CENTURY MANNER

Preparing your own Mustard from Wild Mustard seeds is a laborious process, complicated by problems of identification. Dry, powdered Mustard is available in all markets. This recipe, an adaptation of an old English one, is a delicious, unusual accompaniment to cold meats.

2 tablespoons powdered Mustard (preferably English), 2 tablespoons cider vinegar, 1 teaspoon honey, cinnamon to taste

Combine Mustard and vinegar, adding water if the paste is too thick; it should be the consistency of yogurt. Add honey and mix well. Let stand at least 10 minutes for the flavor to develop. Add cinnamon to taste, starting with ⅛ teaspoon, then add more if you prefer a pronounced cinnamon flavor. Taste cautiously—it is *very hot!*

MUSTARD GREENS

Mustard greens may be readily gathered for eating and are best picked young, so look for them in early summer, or select the new leaves at the tips of the plants. These will be less bitter. Mustard greens are a good source of vitamins A and C; there are 7,000 IU of vitamin A and 97 milligrams of vitamin C per 100-gram serving. The recommended daily allowance is 5,000 IU of vitamin A and approximately 75 milligrams of vitamin C. Raw Mustard leaves can be chopped and added to green salads, or prepared as you would spinach. The fresh flowers can be added to salads also.

Of course if you have a garden, Mustard greens can be grown at home; seeds are carried by most nurseries. Commercially cultivated Mustard greens are available in greengrocers, but they are often quite expensive. Mustard grows wild in the city and is waiting to be gathered and enjoyed.

MUSTARD GREENS WITH SMOKED PORK BUTT

Serves four.

1½ pounds smoked pork butt, 1½ to 2 pounds young Mustard greens, salt to taste

Place pork butt in a pot with enough water just to cover. Bring to a boil, cover the pot, and simmer gently for 1½ hours. Remove pork and keep warm. Wash Mustard greens and remove stems. Add greens to the liquid the pork butt has cooked in. Simmer the greens for 10 to 15 minutes; do not cover the pot or greens will turn brown. When they are done, taste and add salt if needed. Drain greens and place them in the center of a platter. Slice the pork butt and arrange the slices in a circle around the greens. Serve the delicious, vitamin-rich broth in cups on the side, or reserve it for a later use.

Note: Chicory (page 33), Dandelion (page 49), and Lamb's-quarters (page 78) are excellent when prepared in the same manner.

Black Mustard seeds.

The seed pods of Mustard remain after the leaves have fallen off, and are quite attractive on the stems. The colors range from brown and tan to a silvery gray. The seed pods themselves are thin and papery in some species, ribbed and knotted looking in others, and make attractive additions to dried plant arrangements.

PLANTAIN *(Plantago major)*

Folknames: Waybread, Waybroad, White Man's Foot, Englishman's Foot, Cuckoo Bread, Snakeweed, Devil's Shoestring, Common Plantain, Rabbit Ears.

Location: Waste areas, vacant lots, roadsides.

Botanical Description: The ubiquitous Plantain is a low-growing plant. The leaves form a flat rosette and are dark green, oval or spade-shaped, on longish channeled stems. Each leaf is vertically ribbed, and the leaf margins may be wavy or smooth. Plantain has a fairly long blooming season. From June and continuing through November in the eastern United States, numerous flowering spikes grow from the center of the leaf rosette. These are slender, erect, and depending on environmental conditions, anywhere from three to ten inches tall. The flowers themselves are green-tinged and very minute, but they are densely clustered on the flower spike and give it a fluffy, feathery appearance. The seeds of Plantain are small and as densely clustered as the flowers. They are arranged in a complex manner, overlapping each other in a pattern reminiscent of the arrangement of fish scales. The flower structure is complex and fascinating, worthy of close observation with a hand lens.

Common Plantain is an alien, introduced from Europe. It grows throughout the United States and is particularly abundant in the Northeast. A closely related species, Pale Plantain (*Plantago rugellii*) is native. It has red-tinged stalks and the leaves are a brighter green.

Plantain is seen everywhere in the city yet is generally ignored, except by gardeners, who despise it. They consider it an ugly pest and uproot it quickly before it spreads. Plantain is a perennial and sprouts readily from abundantly produced seeds. If cut off at the base, the plants will unhesitatingly send up a fresh crop of leaves and fruiting stalks.

Plantain is ubiquitous because it has learned to survive where most other plants cannot. So highly adaptable is this plant to its environment that one or two will thrive in a narrow crack in the sidewalk or, with more room, spread to form a large carpet of leaves. In this last instance, especially on neglected waste ground or empty lots no one cares about, its presence is an asset; the clumps of broad green leaves are far more attractive than the rubble or otherwise barren dirt would be.

In situations where it is subjected to harsh treatment, Plantain will nevertheless endure, although greatly reduced in size. I have seen miniature plants with leaves no larger than a quarter (in favorable circumstances, these same leaves may be six inches long and three wide) gallantly producing viable seed stalks only two inches high.

Historical Lore, Legends, and Uses: Plantain has been a faithful companion to humankind. Its reputation as a medicinal plant is both ancient and honorable. William Coles noted that the Latin name *Plantago* was derived from the word *planta,* meaning "sole of the foot," alluding to the resemblance between the shape of the leaves and that part of the body. He adds that classical scholars called it *"Plantage, ducto a Planta vocabulo,* the plant of Plants, as it were." He continues, "We in English call it Waybread because it commonly breeds by the wayside. Plantain grows very plentifully . . . and sometimes in gardens, without invitation or welcome though they be as useful as any there."

Plantain was considered particularly valuable as a wound herb, especially useful for staunching blood and joining broken flesh. Two thousand years ago Dioscorides recommended that fresh Plantain leaves be crushed and applied to a wound to stop the bleeding. He suggested other uses as well: the leaves were good to treat burns and other skin inflammations; the fresh juice of the plant healed sores in the mouth and was good to add to eye salves.

Waybread and Waybroad are old Anglo-Saxon names for Plantain, and it was one of their most highly esteemed medicinal plants; the leechbooks mention it frequently and recommend its use for more than twenty different ailments, including one mysterious affliction for which the use of Plantain is advised "in case a man's body be hard." Medicine was combined with magic in a leechdom (prescription) for headache: "Take the roots, tie them to the neck, the headache will go away. For sores on the face, squeeze the Juice of Plantain onto soft wool, lay on. Do this for nine nights; the sores will heal. For scorpion bite, take the root of Waybread, bind it on the man . . . it is believed that it may come to be of good service to him." The Anglo-Saxons used Plantain to treat smallpox, which they called pock disease. To ease the ravages of this terrible scourge, a salve containing Plantain, Yarrow, and Burdock was applied to the skin.

The extent to which the Anglo-Saxons venerated Plantain can best be seen in a poem dedicated to it, one of nine composed in honor of nine sacred plants:

And thou, Waybroad,
Mother of Worts,
Open from eastward,
Mighty within;
Over thee carts creaked,
Over thee Queens rode,
Over thee brides bridalled,
Over thee bulls breathed,
All these thou withstoodest
Venom and vile things
And all the loathly ones,
That through the land rove.

The poem alludes to the remarkable hardiness of Plantain as well as its usefulness in curing illness. The Anglo-Saxons believed many diseases were caused by witchcraft or "venom and vile things that through the land rove." Plantain was an ingredient in an ointment for "flying venom" (an expression probably used to denote epidemics, such as smallpox and bubonic plague). A handful of Waybroad was mixed with honey and other herbs. Butter was melted "three times together," and added, and finally, "Let one sing a Mass over the worts, before they are put together and the salve is wrought up."

Plantain retained its primacy as a wound herb for many centuries. Chaucer referred to it, and Shakespeare mentions it in his plays.

Plantain was useful for a variety of other ailments. To stop vomiting, the patient was fed a cake made with Plantain seeds, egg yolk, and flour. Plantain juice mixed with oil of roses and rubbed on the temples and forehead "helped lunatic persons very much." The roots were good for ordinary headaches.

Plantain did not go unnoticed by adherents of the doctrine of signatures (the philosophy of which is explained in the chapter on Chicory). "It has the signature of the tongue," declared one follower, "which is not only expressed by the outward form, but also by the sinews and veins which run through it. Therefore," he continued, "it avails very much in different diseases of the tongue and inflammation of that and adjacent parts: mouth, gums, throat, etc."

This fanciful notion may have been suggested by the observation that Plantain leaves when torn crosswise expose several white threadlike veins or "sinews," and the shape of the leaf somewhat resembles the shape of the tongue.

In 1824, Dr. Thomas Green described the "remarkable success" achieved by one of his colleagues, who treated liver and chest ailments with a drink prepared from Plantain: the leaves were gathered in the morning after the dew had dried off,

PLANTAIN: *Entire plant,*
approximately life size.

then bruised in a mortar, wrapped in a clean, soft cloth, and placed in hot water for a brief period. After this, the leaves were squeezed and the resulting juice was drunk with wine.

Green was aware of the long history of Plantain as a vulnerary, and he observed: "The common people do now apply the leaves to fresh wounds and sores." In his opinion, however, the seeds were better suited for this purpose because they were quite mucilaginous and therefore more soothing. He added finally, "The root has been recommended for the treatment of intermittent fevers and not undeservedly."

The English colonists brought Plantain to the New World. It seemed literally to spring up in the very footsteps of the European settlers, earning it the names Englishman's Foot and White Man's Foot, as it was called by native Americans. The Delaware Indians called it Path Weed and must have learned of its virtues from contact with settlers, because they used it to heal wounds and treat female disorders.

Plantain was important to the Pennsylvania Dutch, who used it as a medicinal plant and potherb as well. A tea of the seeds was drunk to expel worms, crushed fresh leaves relieved hemorrhoids, and leaves steeped in milk were applied to insect bites.

Today, Plantain is used medicinally by European herbalists. In France, Plantain seeds are obtainable in drugstores. French homeopathic doctors prepare an infusion of the dried leaves for respiratory troubles, and the seeds, soaked in water until soft, are drunk to relieve mild diarrhea and enteritis.

In Britain an extract of the whole plant is used by homeopaths to treat cystitis and hemorrhoids. The 1971 edition of the *British Herbal Pharmacopoeia* describes Plantain as diuretic and antihemorrhagic. Crushed Plantain leaves are recommended for use in staunching the flow of blood in emergency situations when other means of doing so are not available.

In the United States, Plantain was described in the 1917 edition of *Potter's Therapeutics* as "an active hemostatic" (a medicine that stops bleeding). A fluid extract of the plant was sold in drugstores. The pounded fresh leaves applied as a paste or the dry, powdered leaves quickly stopped hemorrhaging. Plantain was used to treat burns, wounds, and Poison Ivy rash.

The *United States Dispensatory* (1955 edition) lists a mucilage-forming preparation whose chief ingredient is material obtained from a species closely related to *Plantago major*, Common Plantain. It is mixed with water and prescribed for treating constipation.

In the course of its journey with humanity, Plantain, as might be expected, has become associated with numerous legends and superstitions. The oldest, most pervasive, and most curious of them all is the supposed influence of Plantain over venomous insects and serpents. It was most likely an ineffective cure, yet the belief persisted for centuries. The Anglo-Saxons said: "For snakebite, take Waybread the wort, rub it into wine and let the patient eat it." Even animals benefited: "A toad being overcharged with the poison of the spider has recourse to the Plantain leaf, which cures him."

It was believed as far back as classical Greek times that the root of Plantain hung about the neck would protect the wearer from the bite of snakes, scorpions, and other harmful creatures, a folk belief that was even transmitted to the native Americans. The Mohegans applied the fresh leaves of Plantain to snakebites to "draw out the poison." The ribbed underside of the leaf was used for this purpose; the smooth upper side had other healing qualities. John Parkinson described a fever remedy containing Plantain. He attributed it to Dioscorides (a classical physician greatly esteemed in Parkinson's day): "If three roots are boiled in wine and taken, it helps the Tertian or three-day fever. Four roots help the Quartian [four-day] fever." Despite the high regard Parkinson had for Dioscorides, he comments: "I hold the number to be fabulous, yet the decoction of them may be effectual."

The Anglo-Saxons used Plantain in the following charm for spoiled milk: "Bind together Waybread and Cress, lay them on the milkpail, and set not the vessel down on earth for seven nights." The charm does not say what will happen to the milk after letting it stand for a week, but it must have been obvious that no charm would make sour milk sweet again. It is possible that spoiled or soured milk left thus long undisturbed would separate into liquid (whey) and solids (curds). Both curds and whey were (and are) edible, even palatable, in this state, and may have been eaten in that form or the curd made into cheese. Perhaps the Anglo-Saxons believed that once milk had "spoiled," the curd-forming process needed the magical assistance of Waybread and cress.

An old Anglo-Saxon legend described the naming of Plantain, or Waybread: a beautiful maiden was married to a knight of great renown. Shortly after the wedding, he was called away to battle. The knight kissed his bride farewell at the roadside and bade her watch for his return. She waited faithfully, but he never came back. Finally she turned into the plant called Waybread, where she remains yet, waiting for her husband to come home.

Plantain, an herb of Venus, was an important element in numerous charms and love divinations. A rite of great antiquity was once practiced in Berwickshire, Scotland: two flowering stalks of Plantain were picked to represent the man and woman in question. The fluffy blossoms were carefully removed, and both stalks were wrapped in a Dock leaf and laid under a stone. The next morning, the stalks were unwrapped. If both stalks bore new blossoms, there would be great love between the man and woman; if only one stalk bore flowers, there would be unrequited love.

Canaries, linnets, and many little wild birds are quite fond of the ripe seeds of Plantain. Bunches of seed spikes used to be hung in birdcages, and wild birds can frequently be observed feasting on the seeds. This habit may well account for the custom of calling Plantain cuckoo bread in Devonshire; it was believed that once every seven years, the Plantain changed into a cuckoo and flew away.

Plantain is much esteemed by modern herbalists. The crushed fresh leaves are believed to relieve the itching and pain of Poison Ivy rash, and preparations of the leaves are used to treat a variety of skin rashes. The root and juice of the plant are used for easing the pain of neuralgia.

Plantain is common in China. The leaves and seeds used to be eaten by the poor, and the plant has been used medicinally since ancient times. The seed is supposed to aid conception, and the leaves are valued as they once were in Europe, for healing minor wounds and nosebleeds.

Plantain seeds may have another as yet unexploited possibility as a food source. In *Weeds and What They Tell Us,* a book about beneficial aspects of "weed" plants published in 1945, the author, Ehrenfried Pfeiffer, mentions that Plantain seeds contain a higher percentage of oil than almost any other seeds, and he describes this oil as being "fine, almost tasteless." Considering the large sums and great effort expended yearly by farmers to eradicate Plantain, perhaps it would be worthwhile to investigate its potential as a culinary or industrial oil.

Suggested Use: The very young leaves, gathered in early April and late May, can be chopped and added to salads. Older leaves are stringy and unpalatable.

POISON IVY
(*Rhus radicans* or *Toxicodendron radicans*)

Folknames: Poison Oak, Poison Vine, Three-leafed Ivy, Markweed, Mercury.

Location: As a vine on trees and fences or a small low shrub under hedges, along walls, or growing in with other plants in parks, gardens, vacant lots, and waste areas.

Botanical Description: Poison Ivy, an all too common plant in the city, may grow as a woody vine or small shrub. Three leaflets grow on single stalks in an alternate pattern on the stems. These leaflets may be quite large, often four or more inches long on the vining form, and are sharply pointed at the tips. They may or may not be shiny. Their edges may be smooth and unnotched, wavy, deeply toothed, or lobed.

Poison Ivy blooms from late May to early July in the Northeast. The tiny yellow-green flowers are borne in loose clusters at the points where the leaf stalks join the main stems. Poison Ivy flowers are sexually mixed on the same inflorescence: that is, three *different and separate* types of flower—male (stamens), female (pistils), and unisexual (stamens and pistils)—occur at the same time in the same flower cluster. The berries are borne in clusters. They are small, round, and covered with a thin, light-gray or greenish papery membrane that splits open to reveal the white fruit.

There is controversy as to whether the vine and shrub are really varieties of one species or two separate and distinct species. In its vinelike form, Poison Ivy is called *Rhus radicans* or *Toxicodendron radicans*. Its leaves tend to be shiny, with unbroken edges. Poison Oak is shrubby, and its leaves, usually lobed and resembling those of an oak tree, may have smooth edges as well. Both forms commonly grow side by side (I have even observed both lobed and smooth leaflets on the same plant) and resemble each other so closely that positive identification can be made only by a specialist. The question is academic, since both inflict an identical toxic reaction.

Poison Ivy is native and widespread throughout the United States. It is highly adaptable, extremely variable, and very hardy. A perennial, it reproduces easily from seeds and underground rootstocks which run horizontally for several feet and are capable of producing a viable new plant at every node. Birds like to eat the berries, and this helps to further spread the plant's growth. *All parts of the plant at all times of the year contain a poisonous sap that can severely blister the skin and mucous membranes.*

In addition to its toxic sap, Poison Ivy has several other traits that are quite sinister. For example, it is an inconspicuous and easily overlooked plant, especially in the spring, when it puts forth small shiny green leaves that are hard to distinguish from the small shiny green leaves characteristic of other (and harmless) plants at that time of the year. In the summer Poison Ivy mingles with and is often camouflaged by other shrubs and vines. It especially likes Blackberry bushes, and in the city I have left unpicked many a luscious cluster of Blackberries because the two plants were so intimately entwined. For the most part, the leaves of Poison Ivy are hardly noticeable and blend all too effectively with the general summer greenery.

In the fall, however, Poison Ivy becomes only too conspicuous, as many hapless victims in search of decorative autumn foliage have discovered. At this time of year, Poison Ivy leaves turn beautiful, striking shades of crimson and orange. Many un-suspecting nature lovers, delightedly collecting huge bunches of the colorful leaves

POISON IVY: (*Above*) *leafy branch,*
(*upper right*) *stalk with berries,*
and (*right*) *leaf variant showing lobed leaflets,*
all approximately life size.

for indoor arrangements, have carried home a severe Poison Ivy rash as well.

The smooth, brown-flecked vine stalks persist throughout the winter, and so, occasionally, do the attractive clusters of gray-green or ivory berries. Having perhaps attracted the eye of a casual stroller, a few inoffensive-looking sprays of berries are picked, brought home, and added to a vase of dried leaves. A few days later, the doctor writes a prescription to ease a painful rash and both doctor and patient are at a loss to account for a case of Poison Ivy appearing mysteriously in the middle of winter.

Poison Ivy is one plant many persons feel they can identify on sight. Even children are familiar with the rhyme: "Leaflets three, let it be." Yet many people continue to suffer as a result of their unwitting contact with it. Approximately one out of every two persons is allergic in some degree to Poison Ivy, and every year in the United States, almost 2 million people develop a reaction from direct or indirect contact. This reaction varies widely, from a mild rash to large, painful blisters that take weeks to heal. In severe cases, the entire system may be affected. The body becomes swollen, the rash spreads all over and may be accompanied by fever, vomiting, and other alarming symptoms. No fatalities have ever been recorded, however, and the illness eventually subsides.

The toxic principles in the sap, as has been noted, are present throughout the year but are strongest in the spring and summer. The sap is released when Poison Ivy is picked or bruised by being rubbed against. An individual can develop a rash simply by stroking a pet that walked or rolled in the plants or from handling clothing or sports and garden equipment that came in contact with it. Reinfection from these articles, if they are not carefully decontaminated, is common. If this were not enough, tiny droplets of sap can be carried in the smoke of burning plants, the pollen from the flowers is airborne, and both are capable of producing severe reactions in highly sensitive persons.

Historical Lore, Legends, and Uses: There is considerable folklore about Poison Ivy, the best-known and most dangerous being the myth that swallowing a fresh leaf will confer immunity. Many persons claim to have done this with no ill effect, or to have picked and crushed the fresh leaves with impunity. This is possible, owing to the great variability of reaction in different people as well as the variable toxicity of the plants themselves, but it is not due to "natural immunity." Such senseless displays of bravado can cause serious blistering of the mouth and throat and should never be done!

Contrary to popular assumption, Poison Ivy rash cannot be "caught" from someone who has it, nor will the blisters, should they ooze or break, transmit the affliction.

To ease the discomfort of Poison Ivy rash, the juice of a plant known variously as Elfcap, Jewelweed, or Touch-me-not (*Impatiens capensis* and *I. biflora*) has been mentioned. More readily available for this purpose in urban areas, however, is the Common Plantain (*Plantago major*). The crushed fresh leaves of this plant have a long history of being applied as a poultice to soothe and help heal the rash.

It has recently been established that toxic substances in the sap of Poison Ivy plants combine immediately upon contact with proteins in the skin to produce the familiar reaction. This would seem to render the well-known "wash with brown soap" remedy largely ineffective, though it may remove excess sap and prevent its transfer to other parts of the body or, via clothing, to other people.

Poison Ivy has been characterized as "clearly one of the worst of all weeds," and it is not generally realized (indeed, it may even come as a shock) that Poison Ivy was once employed medicinally. In our own country, native American people were aware of and had remedies for its irritant properties, but they realized its therapeutic potential as well. Indians of southwestern California made poultices of the crushed fresh leaves and applied them to the skin to get rid of ringworm. The Mohegans of Connecticut, who believed the root of Poison Ivy contained medicinal properties, roasted and crushed it to make a poultice for skin diseases.

In 1798 a physician living in Valenciennes, France, examined a young man who had had a herpes infection on his wrist for many years. After an episode of Poison Ivy inflammation, the herpes disappeared. This seemed to indicate the possible therapeutic value of Poison Ivy as a counterirritant, and the plant was used to treat a variety of chronic and stubborn skin infections for many years, with reportedly excellent results. In the United States as recently as 1917, some authorities strongly recommended it for curing erysipelas, eczema, and herpes zoster (shingles). Several physicians corroborated its value. The use of Poison Ivy in treating these ailments often resulted in serious side effects, and its use was abandoned. Contemporary physicians have safer and more effective ways of managing such diseases.

A nineteenth-century physician named Dr. Kalm performed several courageous experiments with Poison Ivy, using himself as the subject. In one, he bravely squirted the fresh juice into his eye, which fortunately had no ill effect. In another experiment, the doctor covered a volunteer's hand thickly with the juice. "The skin, a few hours after, became as hard as a piece of tanned leather," he reported, "and peeled off afterwards in scales."

Several attempts were made in the nineteenth and early twentieth centuries to procure immunity to Poison Ivy, to be built up gradually by injection or internal medication. At one time, there were several preparations on the market containing extracts of the various poisonous species, alone and in combination. Unfortunately, it was found that in many cases these extracts made the inflammation worse, and their use was discontinued. At present, there is no real cure for the rash; treatment consists of relieving the discomfort until the affliction runs its course. A fluid extract of the fresh leaves is still used by homeopathic doctors who value it in the treatment of various skin disorders.

Poison Ivy is a member of the family *Anacardiaceae,* a group of plants known more commonly as the cashew family. Poison Ivy is thus a relative of the Sumacs, cashew, mango, and pistachio. These last three are highly esteemed (and in this country, expensive) edibles.

When the stems of Poison Ivy are broken or cut, the sap, milky or pale tan at first, becomes black and sticky upon exposure to the air. At one time this sap was a source of black dye. (Before the introduction of synthetic dyes, true black was the most difficult color to obtain with plant materials.) A brown-black ink made from Poison Ivy sap is so indelible on fabric that it cannot be removed; it actually gets darker the more it is washed.

Suggested Use: None.

POKEWEED *(Phytolacca americana)*

Folknames: Poke, Skoke, Pocan, Coakum, Pokeberry, Crowberry, Bear-Grape, Pigeonberry, Inkberry, American Spinach, Cancer Root.

Location: Along fences, in parks and waste places, at the edge of open areas, often found where the ground has been recently disturbed.

Botanical Description: Pokeweed, a treelike shrub, grows from four to ten feet high. The stems are stout and hollow; one or several grow from the large, fleshy root, and as the plant matures, they develop a beautiful rose-red color. The leaves are light green and oval, with smooth edges and pointed tips. They may be several inches long.

Pokeweed is in bloom during July and August in the eastern United States. The flowering stems are long, drooping, and paired with a leaf on the main stalks. The individual flowers are white, dainty, and very small. The exquisite symmetry of their structure and growth should be closely observed to be really appreciated. As the berries develop, their stems become rose-colored and the berries change from green to glossy deep purple.

Pokeweed is such an attractive plant that one writer eloquently observed that "its large rich leaves and beautiful clusters of purple berries, often mingled upon the same branch with the green unripe fruit and the flowers still in bloom, render it one of the most striking of our native plants."

Pokeweed is indigenous to North America and common throughout the eastern United States. Several species of the Pokeweed family (*Phytolaccaceae*) grow in Mediterranean countries, but only one species, *Phytolacca americana,* is found in this country. Pokeweed is a perennial and reproduces from seeds.

Pokeweed contains several physiologically active principles. The toxic substances are most highly concentrated in the root, but the mature leaves, stems, and purple berries are also considered poisonous. The young shoots, properly prepared, are edible and delicious.

Historical Lore, Legends, and Uses: The name *Phytolacca* is derived from the Greek *phytos,* meaning "plant" and the French *lac* and Italian *lacca,* meaning "crimson" or "red." Poke, or Pokeweed, the common American folknames, are derived from the languages of native American peoples. They variously called it Skoke, Pocan, and Coakum, words used to denote a plant that yielded a red or yellow dye. The juice of ripe Pokeberries was employed by various native American tribes to decorate splint baskets.

As a medicinal plant, potherb, and dye source, Pokeweed was well known by and extremely valuable to native Americans, especially to tribes inhabiting the Southeast, where the plant was most common. The Lenape chopped the root, poured boiling water over it, and prepared a liniment to reduce swellings. To reduce fever, they bound the fresh roots to the hands and feet. Other tribes made a purge from the juice of the root.

The Delaware considered the roasted mashed root of Pokeweed (they called it Purple Berry) an excellent blood purifier and stimulant. They were well aware of the toxic properties of Poke root, and only very small doses were administered. Poke root was combined with Bittersweet (*Solanum dulcamara*) by other tribes and used as an ointment for chronic sores, and the Pamunkey of Virginia treated rheumatism with preparations of the boiled berries. The Mohegans of Connecticut ate the young shoots in the spring (as did many other native Americans) and used poultices of the mashed ripe berries to relieve sore breasts of nursing mothers.

Beautiful colors ranging from delicate rose-pink to deep purple-red have been obtained from the juice of ripe Pokeberries, but unfortunately these colors are not fast on textiles, and fade rapidly. The Pennsylvania Dutch are known to have made an ink from the plant, and one nineteenth-century writer mentions that a permanent blue textile dye could be obtained with stale urine as the mordant, but we no longer have the formulas for either. (Unappetizing as it sounds, urine was commonly employed to fix the color of dyes obtained from plant material. There were two excellent reasons for this: it gave dependable results and was readily available!) George Caleb Bingham, one of our finest early American painters, used the juice of ripe Pokeberries as a pigment.

The vintners of Portugal discovered a profitable and ingenious use for Pokeweed. The juice of the berries gave color to their famous ruby port, a fact not generally known. Perhaps greed prompted its eventual use as an adulterant; this gave the wine so disagreeable a taste that complaints eventually reached the government and the practice was banned. It did not actually cease, however, until the authorities ordered the plants cut down before berries were produced.

Pokeberries are relished by wild birds, and feeding experiments with domestic fowl caused no ill effects, but in some instances the birds were observed staggering about as if drunk! They were unfit to eat, however, because the berries made their flesh extremely bitter.

While it has long been known that the roots of Pokeweed are very poisonous, the toxicity of the ripe berries has been the subject of continued controversy. They are the most conspicuous part of the plant and therefore quite likely to be eaten.

Some authorities claim that children have been fatally poisoned by eating the berries; others maintain the ripe berries are harmless, and that deaths attributed to this source are not conclusive. A few writers have said that cooked Pokeberries are edible and delicious and have been used by country people for jam and pie fillings, without harm. But there is really little reason for eating a substance of questionable safety, so Pokeberries in any form are best avoided.

Toxicity is *not* a problem where the edible young shoots are concerned. Native Americans eagerly awaited their arrival in the spring and probably recommended this potherb to the early settlers, who quickly learned to love it. Poke greens have long been a favorite in rural areas. When properly prepared, they are perfectly harmless and very delicious. The Pennsylvania Dutch sell quantities of Pokeweed shoots every spring in the markets in and around Lancaster, Pennsylvania.

Pokeweed was listed officially in the *United States Pharmacopeia* for nearly one hundred years, from 1820 to 1916, and a fluid extract of the dried root was prescribed for a variety of ailments. During the early 1900s, it was a major ingredient in a popular over-the-counter obesity remedy, Phytoline, taken six times a day, before and after each meal. Its efficacy, however, would seem questionable.

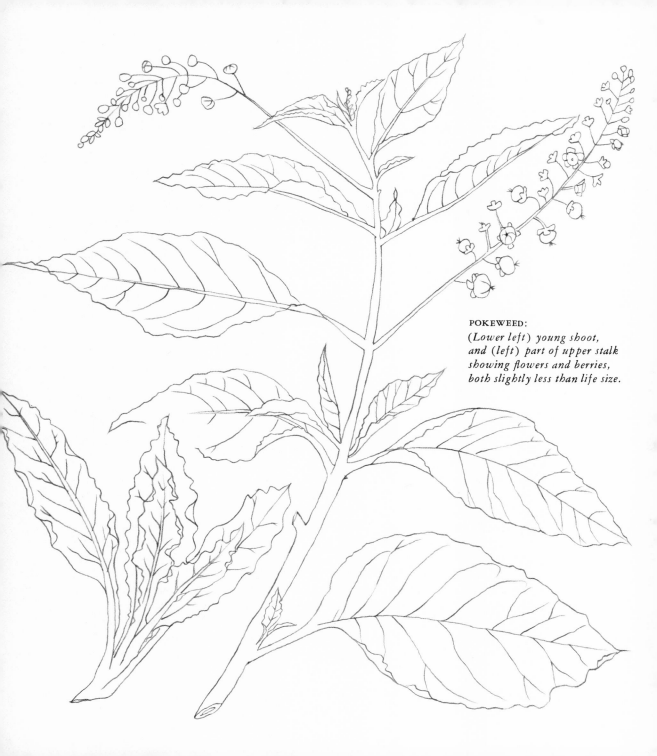

POKEWEED:
(*Lower left*) *young shoot,
and* (*left*) *part of upper stalk
showing flowers and berries,
both slightly less than life size.*

Pokeweed was employed by old-time "herb doctors" in many a folk remedy. A "cancer cure" was prepared by mixing the juice of the leaves or root with gunpowder, and in the Ozark Mountains, Poke was a famous remedy for a variety of parasitic skin afflictions collectively known as "the itch." The root was boiled into a thick paste and reputed to work very well, but was quite painful when applied.

Old Chinese herbals describe two varieties of Pokeweed. One, believed by some authorities to be *Phytolacca americana,* had a white root and white flowers, and a drug prepared from its root was used for treating dropsy and infected wounds. Its white flowers, called *Chang hua,* were employed to treat apoplexy. The other Pokeweed variety had red flowers and a red root and was considered poisonous.

China, Pokeweed, and the United States were once involved in a curious relationship. At one time, ginseng root (*Panax quinquefoilus*), or sang, as it was commonly called, was gathered in this country for export to China, where it was—and still is—highly valued, and consequently brought a very handsome price. The plants were so ruthlessly collected for this purpose that they became extinct in some areas. To keep the supply and money going, unscrupulous collectors began to pass off dried Pokeweed roots as "real sang." They were quite successful until the practice was discovered and stopped.

Pokeweed even had a brief and amusing role in United States politics. During the presidential campaign of James K. Polk, his supporters made it a point to wear a boutonniere of Poke leaves in their lapels!

A bizarre anecdote concerning Pokeweed is related by Richard Folkhard in *Plant Lore and Legends:*

A species of *Phytolacca* was found in Nicaragua in 1876 . . . and named *Phytolacca electrica.* [It] may well be called the electrifying plant. The discoverer, while gathering a bunch, experienced a veritable electric shock. Experimenting with a compass, he found the needle was agitated at a distance of eight paces and became more so the nearer he approached, the action changing to a rapid gyratory motion when he finally placed the compass in the midst of the shrub. There was nothing in the soil to account for what might be termed the "shocking" proclivities of the shrub, which are slight in the night-time, becoming gradually intensified until about two o'clock p.m. In stormy weather, the intensity of action is increased and the plant presents a withered appearance until the fall of rain. Neither insect nor bird was seen to approach this terrible shrub.

This intriguing story has all the earmarks of a hoax, or at least an exaggeration. Search as I might, I was unable to find any further mention of it in the literature, nor is the "scientific" name *Phytolacca electrica* anywhere recorded. Alas.

Pokeweed has recently attracted the attention of researchers and scientists involved with disease control. In certain parts of Ethiopia, a few years ago, large numbers of dead snails were seen in the small waterways downstream from where clothing was washed. Investigators found that the native women laundered their clothes with suds-producing berries of a native plant, *Phytolacca dodencandra.* Immediately upstream, however, the snails were alive and quite unaffected, suggesting that some agent in this "vegetable soap" was toxic to the snails. The snails in question are important because they are carriers of bilharziasis, a severely debilitating parasitic disease widespread and endemic in Africa and other tropical areas throughout the world.

American Pokeweed, *Phytolacca americana,* exhibits the same chemical characteristics as the Ethiopian species (to which it is closely related), to an even greater degree. From it researchers hope to develop a molluscicide that will help eradicate this disease, the control of which has proved very difficult. So far, experiments have been promising, and it has been further demonstrated that the molluscicidal potency of the material is not affected by the presence of soil or vegetable matter, both important factors when the molluscicide is used (as it must be to be effective) under natural conditions.

Once a folk remedy for cancer, Pokeweed is currently being studied by scientists at the National Institutes of Health. Investigators have reported finding a mitogenic (i.e., initiating mitosis) substance in Pokeweed that may prove useful in cancer research and treatment.

Suggested Uses: The young green shoots of Pokeweed are a popular potherb in many rural areas of the United States. They are gathered in the spring and often sold in local markets. Pokeweed is common in some urban areas as well. If you are not familiar with the plant, look for Pokeweed in October, when the plant, with its magenta stems and long, drooping clusters of shiny purple berries, is both conspicuous and beautiful to behold. Note the location of the plants so that in spring you can return to gather young green shoots for the table. When properly prepared, Pokeweed shoots are excellent, truly one of the finest wild greens, and nutritious, too. Cooked Poke shoots contain 8,700 IU vitamin A and 82 milligrams vitamin C per 100-gram serving. The recommended daily allowance is 5,000 IU vitamin A and about 75 milligrams vitamin C.

Gather the shoots when they are no more than six or eight inches high. Put them into a saucepan with water to cover and cook them for 10 minutes. Pour this water off and add fresh boiling water to cover. Cook 5 minutes more. Discard this second water, drain well, and serve hot with plenty of melted butter and seasoning to taste.

Shades of deep rose to light pink can be obtained from ripe Pokeweed berries. See the Appendix, page 172, for instructions on preparing dye.

PRICKLY LETTUCE *(Lactua scariola)*

Folknames: Compass Plant, Wild Lettuce, Opium Lettuce.

Location: Along sidewalks and embankments, in vacant lots and waste areas.

Botanical Description: Prickly Lettuce is a tall plant, growing between two and four feet high, of a beautiful blue-green color with a whitish bloom. The stem is erect and branched. The lower leaves are quite large, and they may be unlobed or deeply notched, with toothed margins. The bases of these leaves clasp the stem. The midribs are set with numerous weak prickles, and these are present on the lower stems as well. The leaves on the upper stems are much smaller and delicately arched. Their bases tightly clasp the stems. A bitter, strong-smelling, sticky white sap flows abundantly when the plant is broken. Prickly Lettuce flowers during July and August in the Northeast. The flowers are light yellow and resemble miniature Dandelions. The seed heads are white and fluffy.

Prickly Lettuce is ubiquitous in the city, and practically everyone has seen it. I have found plants that were only eight or so inches tall blossoming and seeding in seemingly intolerable situations. This remarkable hardiness, noted elsewhere in regard to other plants, is a key to their successful survival.

An unusual feature of Prickly Lettuce is the orientation of its leaves. In sunny areas, the leaves twist at the base and face east to west, with their edges pointing due north and south. This explains the name Compass Plant, by which it has been known since classical times.

Prickly Lettuce is an annual and grows readily from easily dispersed seeds. This particular plant is an alien introduced from Europe, but there are several related species. It is widespread throughout the United States.

Wild Lettuce (*Lactua canadensis*) is a native species similar in appearance to Prickly Lettuce. It is smooth, lacks prickles, and has a whitish bloom. Rather common in urban areas, it often grows side by side with Prickly Lettuce.

Historical Lore, Legends, and Uses: Prickly Lettuce, the ancestor of our familiar salad plant, has been cultivated for centuries. The Greeks and Romans loved to eat it, and it was a popular "sallet" green during the Middle Ages. There are now literally hundreds of varieties, and lettuce is a favorite with home gardeners.

Most people who grow lettuce harvest it when it is very young, but if a plant is allowed to bolt (go to seed), the resemblance between table lettuce and its wild

PRICKLY LETTUCE:
(*Left*) *flowering portion of upper stem,*
and (*far left*) *young shoot,*
both approximately life size.

ancestor is soon apparent. The two will physically resemble each other, and the stems and leaves of garden lettuce will contain a milky juice and will taste bitter. Allowing a garden lettuce plant to go to seed, and then comparing it with a Wild Lettuce plant in taste and appearance is an experiment children might like to do.

Prickly Lettuce was called *Lactucarium,* from the Latin word meaning "milky." The leaves were reputedly thirst quenching, and the sap was believed to have a soothing, sleep-producing effect and special curative powers. The Roman Emperor Augustus became convinced that a medicine of lettuce saved his life when he lay near death, and in gratitude and admiration he caused a statue and altar to be erected in the plant's honor.

In the United States, various native American peoples prepared a tea from Prickly Lettuce. It was given to nursing mothers, to promote lactation. This had been an old use of the plant in Europe and was perhaps learned from the settlers. The Menominees used the juice of the plant to relieve Poison Ivy.

During the nineteenth century, American physicians believed the juice of Prickly Lettuce to be an excellent, even preferable, substitute for opium. It was said to induce sleep without the deleterious side effects of opium because it was not narcotic. An English doctor of the same period was of the opinion that true opium was dangerous in inexperienced hands (and there was little regulation of its use at that time). He advised instead a syrup of Wild Lettuce: "It will ease pain and will cause sleep in the manner of that foreign drug without any ill consequences from it."

The sap of Prickly Lettuce was collected for medicinal use in the following manner: the tops of the plants were cut off and the sap collected in small china cups. When the sap was dry, the cups were warmed and tapped to release the contents, now hardened and gumlike. These were cut into quarters and allowed to dry further. At one time, the United States was a customer for this substance. It was imported from abroad and frequently was found as an adulterant of true opium.

Wild Lettuce and garden lettuce have been the subject of numerous folk beliefs and superstitions, a few of which contradict each other. The Romans believed lettuce excited love and passion and warded off drunkenness. It was eaten at banquets for the latter reason and was a prime ingredient in love charms and potions during the Middle Ages for the former. The leaves eaten in salads and the seeds drunk in wine were believed to provoke lust and aid conception.

In medieval times, however, it was thought that evil spirits hid in lettuce beds. These spirits were especially dangerous to pregnant women and newborn infants, and pregnant women were not supposed to eat it. This ancient belief was remembered in an old saying: "Too much lettuce in the garden prevents the young wife's bearing." According to others, however, Wild Lettuce was to be drunk by those "troubled with unclean dreams" and the plant was "good against the rages of venery."

In England, as recently as 1951, the British *Daily Mirror* printed a series of letters on the subject of lettuce and fertility. A writer asked if it were true that eating lettuce was bad for brides. In answer, a woman responded that after being childless for a number of years, she was advised by a specialist to eat plenty of lettuce and to give her husband some too; in less than six months she was pregnant.

Some believed eating quantities of lettuce benefited weak eyes; others claimed that eating too much harmed eyesight. In any case, to dream of it was generally accounted a bad omen, and foretold trouble and difficulties ahead.

Prickly Lettuce was used in the United States as a cough sedative, and a tea prepared from the leaves was considered good to quiet nervous irritation. This effect is now thought to have been psychological, rather than due to any active medicinal principle; supposed medicinal effects of the plant are believed to have been based on the superficial resemblance of its milky sap to opium.

Suggested Uses: The very young leaves of Prickly and Wild Lettuce may be collected in spring and, if not too bitter, added to salads.

COOKED PRICKLY LETTUCE

As a potherb, pour boiling water over tender young leaves to cover, boil for 5 minutes, and repeat the procedure. Drain. Add butter to taste, let it melt, then add a few spoonfuls of cider vinegar. Toss to coat leaves with butter and vinegar. Serve hot.

QUEEN ANNE'S LACE *(Daucus carota)*

Folknames: Wild Carrot, Bee's Nest, Bird's Nest, Laceflower, Devil's Plague.

Location: Dry, open sunny areas—roadsides, vacant lots, and waste places generally.

Botanical Description: Queen Anne's Lace is a biennial, reproducing by seeds. During the first year, only a rosette of finely divided leaves grows from the taproot. This root is white and spindly, with a characteristic carrotlike odor. Unlike carrot, it is tough and stringy and has an unpleasant, acrid taste. The second year, tall (often up to two feet) branched stems are produced. The leaves on the lower stem are large, much divided, and feathery; the upper leaves are less divided and much smaller.

In the Northeast, flowers start to appear in July and the plants may continue blooming well into October. The flower heads are white, disk-shaped, and flat, often with a tiny purple floret in the center. The individual flower disk is actually composed of hundreds of minute, separate white flowers, those on the outer rim of the disk being larger and more irregularly shaped than the inner flowers. The individual flower-bearing stalks all arise from the same point at the end of the main stem like rays or umbrella spokes. Each small stalk divides again in the same manner; and the tiny individual flowers grow from this small stalk.

As the seeds ripen, the flowers fall and the disk contracts. The outer stalks become longer and curve over and around the center. The head becomes bowl-shaped, turning from green to brown. At this stage it greatly resembles a small bird's nest, hence the popular folkname.

Queen Anne's Lace grows throughout most of the United States. A native of Eurasia and common in Europe, Queen Anne's Lace arrived in America with the early settlers.

Queen Anne's Lace is harmless, but it is related to and resembles a few extremely dangerous poisonous plants. Among these are water hemlock (*Cicuta maculata*), poison hemlock (*Conium maculatum*), and fool's parsley (*Aethusa cynapium*). Water hemlock and poison hemlock have purple-spotted stems, while the stems of Queen Anne's Lace are unspotted. Fool's parsley has an extremely unpleasant smell, whereas Queen Anne's Lace has a characteristic carrotlike aroma. In addition, Queen Anne's Lace forms a distinctive cuplike shape when it goes to seed.

Historical Lore, Legends, and Uses: Our familiar garden carrot is a direct descendant of the wild Queen Anne's Lace. The two share the same scientific name, the garden carrot being called specifically *Daucus carota,* variety *sativa.* Cultivated carrots, if left in the ground to go to seed, will develop flower heads similar to Queen Anne's Lace. If the plants are pulled up when the seeds are ripe, and seeds planted from those whose roots most resemble the Queen Anne's Lace root (it may take a few generations of planting such seeds for this difference to become more pronounced), one will eventually end up with a "wild" plant again. If you have a garden you might want to perform this experiment yourself.

Queen Anne's Lace and cultivated carrot were well known to the classical Greek and Roman naturalists. There was some confusion about the plants, because different writers gave them different names. In addition, carrots were often described and

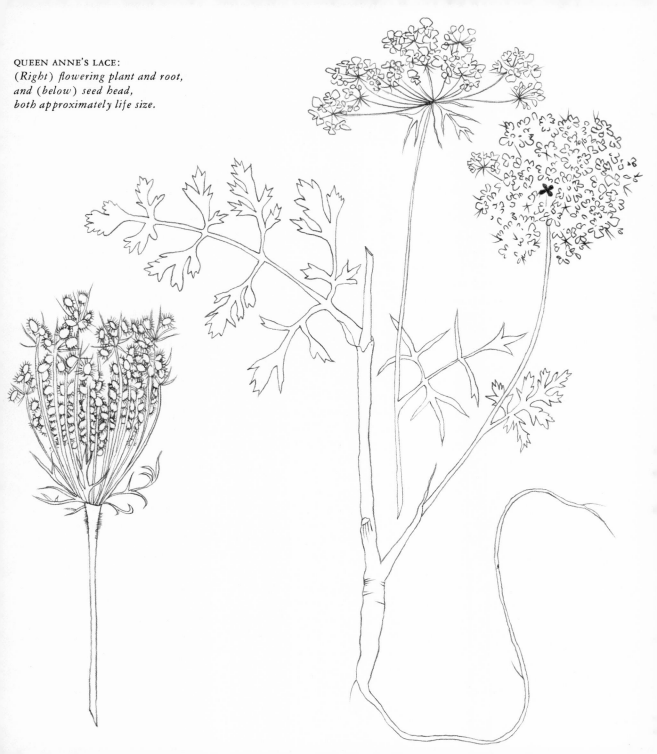

QUEEN ANNE'S LACE:
(*Right*) *flowering plant and root,*
and (*below*) *seed head,*
both approximately life size.

classified with skirret and parsnip, root vegetables somewhat resembling carrot.

The physician Galen used the name *Daucus* to differentiate the carrot from the parsnip, and the name *carota,* as actually applied to the carrot we know, has been found in a cookbook written about A.D. 230 by one Apicus Caelius. The two names *Daucus carota* were officially adopted in the eighteenth century by Linnaeus.

Most herbalists of the sixteenth and seventeenth centuries recognized the relationship between the two types of carrot, referring to them as the "wild" and the "tame." The "tame" carrot had become a popular food by the seventeenth century. Its "pleasant, sweet taste" was very welcome, and carrots were "always sown in gardens or fields chosen out for that purpose."

The great Tudor herbalist-physician William Turner discussed wild and cultivated carrots at great length. In his day (1551) carrots had already become a well-liked food. They were brought to England via Holland, where improved varieties had been developed. Turner noted that classical writers described "carrots" as having an assortment of root colors ranging from white to greenish, yellow, orange, and even black! These old writers, he believed, apparently grouped together such root vegetables as salsify, parsnip, carrot, and radish (one variety of which has a black root—actually black outside and white inside).

The pretty folkname Queen Anne's Lace, by which the wild plant is so familiarly known, has been traced to Queen Anne, wife of James I of England. At one time it was fashionable for the Queen and ladies of her court to dress their hair with the lacy leaves. This is mentioned by Thomas Green, a nineteenth-century writer. "The females of this *polished* age," he wrote in 1824, "will smile at the simplicity of ancient times when they are informed that the autumnal beauty of the leaves [of Queen Anne's Lace] allured many gentlewomen of a former age oftentimes to stick them in their hats or pin on their arms instead of feathers."

Queen Anne's Lace has a long medicinal history. The seeds contain an aromatic and pungent-tasting green oil and were believed powerfully diuretic; they were used to treat urinary and menstrual disorders. The seeds were powdered and drunk in wine to heal spider bites, and this drink was considered, in the words of a medieval observer, "profitable to resist venom or poison, and the Pestilence." Mixed with honey, the seeds were good for coughs. One herbalist praised the seeds for being "especially good and powerful to expel wind and good to ease the gripings and torments of the belly." This same gentleman described the pleasing flavor of the cultivated carrot, but recommended moderate use of it as a food: "Carrots are somewhat windy," he cautioned, "whence you may observe that the roots and seeds of many things are not imbued with the same qualities."

The roots of Queen Anne's Lace have had an ancient reputation of provoking lust in men and carnal love between men and women, and in the East they were considered a powerful aphrodisiac. It was believed that the seeds "provoked venery and stirred up bodily lust." These side effects notwithstanding, the seeds were prescribed for relieving "stitches" in the side and easing difficult births, and were said to impart a pleasant and agreeable flavor to beer if added when the brew was working in the vat.

In the United States several Algonquin Indian tribes made a tea for treating diabetes from Queen Anne's Lace blossoms steeped in hot water. The Pennsylvania Dutch called Queen Anne's Lace *Gehl Reewah* and made a tea from it that they, too, used for diabetes.

In eighteenth-century England, Queen Anne's Lace was popularly employed medicinally. A physician complained, "The shops are often supplied with old seeds of Garden Carrot instead of the fresh seeds of the wild plant. This is one of the ways efficacious medicines are brought into disrepute. But the remedy consists of everyone gathering the wild seeds fresh for himself."

As recently as the mid-nineteenth century, American physicians praised Queen Anne's Lace for its medicinal virtues. The root was extolled as "a powerful stimulant that helps colds, removes obstructions, and promotes the menses and secretions in general."

There is a peculiar (and in some areas still extant) superstition associated with Queen Anne's Lace. In some plants, there is a tiny purple floret growing in the center of the otherwise white flower head. If eaten, this little floret is said to cure epilepsy.

Suggested Uses: Queen Anne's Lace makes an elegant indoor decoration. The entire plant, root included, should be dug carefully out of the ground just before it comes into bloom and placed in a vase of water. The flower buds will continue to open and the plant will last quite a long time, providing a living, continually changing flower arrangement of singular beauty.

The brown seed heads gathered in the fall are attractive in dried plant arrangements, and the seeds are especially nice for flavoring soups and stews. Try substituting them in any recipe calling for caraway or dill seed.

To prepare seeds for tea and seasoning: Gather seed heads in the fall when they look like a small bird's nest and are dry and brown. Strip off as much stalk as possible. Working over a bowl or pail, rub the seeds between the palms in a light breeze so that the bits of chaff will blow away. They can be cleaned further by being rubbed in a strainer. Store seeds whole, away from light and heat. Crush in a mortar or grind in a blender as needed.

A pleasant aromatic tea, useful for relieving mild stomach upsets and flatulence, can be prepared with the dried seeds of Queen Anne's Lace.

Crush 1 tablespoon of seeds with a mortar and pestle, pour 1 cup of boiling water over them, and let steep 5 minutes. Strain and sweeten with honey.

Queen Anne's Lace is a good source of natural dye. Shades of yellow to pale yellow-green can be obtained from the whole plant. See the Appendix, page 172, for instructions on preparing dye.

RAGWEED *(Ambrosia artemisiifolia, A. trifida)*

Folknames: Common Ragweed, Giant Ragweed, Crownweed, Horseweed, Black-weed, Hogweed, Mayweed, Wild Tansy, Roman Wormwood, Carrotweed.

Location: Waste areas, vacant lots, roadsides, throughout.

Botanical Description: Common Ragweed and Giant Ragweed are among the most ubiquitous plants in the city. The former, *Ambrosia artemisiifolia,* often grows to a height of four feet. The stems are somewhat downy, erect, and one or more may grow from the root. The leaves are finely divided and lacy. They resemble the leaves of the *Artemisias* (such as Mugwort), and it is from this similarity that the species name *artemisiifolia* is derived.

Giant Ragweed (*Ambrosia trifida*) is a much larger plant, capable of growing fifteen feet tall. The leaves may be larger than a man's hand, and they are three-lobed and pointed at the tips. It is not unusual to find both species growing side by side.

Ragweed pollen, a major cause of suffering to hayfever victims, is produced in huge quantities in the Northeast from June to September by the *male flowers*. These are borne on long, slender spikes at the ends of the stems and are inconspicuous, minute, and densely clustered. The female flowers are separate, far less numerous than the male flowers, and are borne on almost nonexistent stems at the joint of leaf and stalk.

All Ragweeds are shallow-rooted annuals, indigenous to North America and widespread throughout the United States.

The pollen generated by Ragweed is the most abundant and toxic of all the airborne allergy producers. Most of the eastern United States has a large count, especially in years when the climate is favorable: warm, rainy weather in spring and a dry, sunny summer and fall.

The Great Plains area between the Rocky Mountains and the Appalachians must be credited, albeit dubiously, with creating conditions that expedite the production of Ragweed pollen in truly amazing quantities. This is due to the fact that after the huge fields of grain are harvested, thousands of acres of America's Breadbasket are replaced with luxurious stands of Ragweed. As summer advances and the plants mature, pollen is shed by the billions of grains and dispersed by the gentlest of breezes. Since 90 million grains of pollen weigh only about one gram, this near-weightlessness enables them to travel great distances.

Ragweed is difficult to eradicate when it is left to set seeds. Seeds are produced by the thousands on a single plant; they are easily spread and grow anywhere—I have seen Ragweed plants growing out of minute cracks in the sidewalk. Common Ragweed in particular has no difficulty adapting its size to its environment. A six-inch-tall miniature of its more usual three- or four-foot self will nevertheless flower, release pollen, and set seeds.

Strangely enough, however, Ragweed is extraordinarily vulnerable to loss of moisture. It withers literally within seconds of being uprooted or picked, and broken-off stems will not revive, even after being placed in water. I expect it could be readily controlled if plants were pulled up before they set seeds. This ought not be difficult, since even very large plants have shallow roots and are readily uprooted.

Historical Lore, Legends, and Uses: Ambrosia, as most of us know, was the name classical Greeks gave to the food of the gods, the eating of which conferred immortality. This ambrosia is believed to have been a fragrant herb, but it has never been positively identified. I wondered how our Ragweeds received this poetic name; they are New World species and were unknown in Europe before the sixteenth century. There is a connection, however, and it is a fascinating one.

The ambrosia of the gods was said to be fragrant. There is a plant, native to Europe and Eurasia, that is highly aromatic and was long known as sea ambrosia. When Linnaeus undertook the classification of plants in the eighteenth century, he named this plant *Ambrosia maritima.*

Eventually, the New World Ragweeds were found to be members of this genus and received the generic name *Ambrosia.* The unexpected surprise, however, is that our related native Ragweeds are said to be fragrant as well! I have not yet had the dubious pleasure of verifying this personally, but one day I will stride boldly up to the pollen-laden flowers and, in the name of science and research, inhale!

RAGWEED: (*Right*) *portion of stalk of Giant Ragweed with leaf and seeds, and (far right) stalk of Common Ragweed with leaves and pollen-bearing flowers.*

It must seem impossible to imagine that such a misery-producing plant should have redeeming, much less beneficial, properties. Yet this is the case, though few people are aware of it.

Native Americans were very familiar with Ragweed. Several tribes used fiber from the stems of Giant Ragweed to make rope. The Nanticokes of Virginia considered it a medicinal plant. They prepared a strong laxative from the roots and applied poultices of the fresh green plant to wounds to prevent blood poisoning. These virtues may have been taught to the early settlers, because American physicians of the eighteenth and early nineteenth centuries credited Ragweed with having valuable antiseptic and emollient properties.

Ragweed is discussed by Vance Randolph, a writer who lived in the Ozarks during the 1920s and 1930s and collected a great deal of regional folklore. Several of the old "granny women" told him that the fresh leaves of Ragweed steeped in cold water made an excellent (and widely popular) diarrhea cure. He tried this himself and reported that it worked "like magic." Randolph also described a "hiccough" cure he'd heard about; it consisted of tansy, Ragweed, and whiskey. According to the elderly herbalist who was the source of this remedy, "It agrees with me fine. I take it every day and I ain't had the hiccoughs but once in fourteen years!"

Ragweed seeds are relished by game fowl and are a valuable winter source of food for many wild birds, who are frequently seen eating seeds from dried branches of Ragweed that protrude from the frozen or snow-covered ground.

If Ragweed plants are plowed under or added to compost heaps (this must be done only with plants that have *not set seeds*), they make an excellent, free, green manure and are first-rate soil conditioners.

It has been said that the oil-rich seeds of Ragweed contain an extractable oil supposedly similar to soybean oil, and might be of value for industrial use or even as a food.

Crushed fresh Ragweed leaves are believed to ease the pain and itching of Poison Ivy. Modern herbalists employ preparations of Ragweed to allay the after-effects of quinine, and strong infusions are used externally for treating leukorrhea.

Giant Ragweed leaf,
much reduced.

SHEPHERD'S-PURSE

(Capsella bursa-pastoris)

Folknames: Shepherd's-bag, Caseweed, Witches'-pouches, Toywort, Lady's-purse, Rattlepouch, Clappedpouch, Sanguinary, Mother's Heart.

Location: Paths, roadsides, grassy strips, waste areas.

Botanical Description: Shepherd's-purse is a small, inconspicuous plant, usually only six or eight inches tall. The leaves grow in a rosette and are rather small. They are quite variable in shape, however. Even on the same plant, some will be coarsely toothed and others irregularly lobed. Slender stems, occasionally branched, grow up from the center of the rosette. These stems have very small, clasping, arrow-shaped leaves. The flowers are white and very minute. The seeds are contained in distinctive heart-shaped pods arranged alternately along the stems. In the eastern half of the United States, Shepherd's-purse blooms continuously from March to December. It is an annual and reproduces from seeds. Having accompanied the European colonists on all their migrations, Shepherd's-purse has become established worldwide. In the United States, where it is an alien, it has become naturalized and grows everywhere.

Shepherd's-purse has been blessed by nature with an extraordinary ability to survive. The tiny white flowers are almost continuously in bloom, are self-pollinated before they even open, and do not depend on the vagaries of outside pollinating agents to ensure reproduction. This makes for an uninterrupted supply of seeds, which themselves can lie dormant in the soil for thirty-five years and more, sprouting when conditions are favorable. Shepherd's-purse flourishes in any and all environments. If the situation is advantageous, plants may reach a height of two feet. When its height is restricted by harsh treatment or poor conditions, Shepherd's-purse sets viable seeds from plants only a few inches tall. The plants are extremely hardy, being among the last to die in winter and the first to sprout in the spring.

Historical Lore, Legends, and Uses: Shepherd's-purse (the most common folk-name) and other similar names for it are derived from the resemblance of the heart-shaped seed pods to the heart-shaped leather pouches in which, for centuries, shepherds used to carry their food. Parkinson believed Shepherd's-purse had no classical Greek name, but he noted that all the writers called it *Capsella bursa-pastoris.* In Latin, *bursa* means "bag" and *pastoris,* "pertaining to a shepherd." This ancient name was retained as the official scientific name for the plant. Lady's-purse, another

SHEPHERD'S PURSE:
(*Left*) *plant with seed stalk,*
approximately life size,
and (far left) detail of seed,
about two-thirds larger
than life size.

common name, is a shortened form of Our Lady's Purse. Any plant bearing the word "Lady" in its name was in some manner associated with the Virgin Mary. Shepherd's-purse was long associated with the shepherds of Bethlehem; perhaps it was by this connection that the plant obtained Mary's name.

Shepherd's-purse has had a curious association with lepers, and this too is reflected in its folknames. During the Middle Ages, these unfortunates were shunned by all and condemned to spend their lives wandering from place to place. They were given leave to beg for alms, however, and they used to beg at the roadsides in all the countries of Europe. To attract the attention of passers-by, they had a bell, or clapper, and a bowl attached to a long stick. The alms they collected were put into leather pouches tied at the waist. The names Clappedpouch, which is Irish, and Rattlepouch, which is English, date back to the time when these pathetic outcasts, once quite numerous, begged for their meager sustenance at the side of the road.

As a medicinal herb, Shepherd's-purse has been well known for many centuries. It was considered of particular value in controlling all kinds of internal bleeding and was used for spitting of blood, bloody urine, and nosebleeds. To cure a nosebleed, it was recommended that the plant be held in the hands, bound around the neck, or even better, drunk with Plantain water or red wine.

The seeds of Shepherd's-purse were considered astringent and binding; the leaves were used for curing scurvy and reducing fevers. The juice of the plant helped running ears and cured shortness of breath. The Anglo-Saxons said that "the ooze of this wort, well wrung out and a cupful drunk, removes all evil gatherings of the inwards."

The plant was known to the Amish as *Bochseckel* and *Deschelgraut*. The young plants were eaten as greens and were used as a poultice for wounds. A tea for treating fevers and kidney ailments was prepared from the seeds.

Shepherd's-purse was listed in the 1955 edition of the *United States Dispensatory;* a fluid extract of the whole plant was reported to have a stimulating effect on the uterine muscles, arrested hemorrhage, and lowered blood pressure. During World War I, when drugs to control bleeding were not available (they came from Germany), the British prepared an extract of Shepherd's-purse. It was a successful substitute. Shepherd's-purse is no longer official; however, modern homeopathic doctors do use preparations of the fresh plant to regulate menstruation, and they consider it to be an excellent spring tonic for the system.

The Chinese knew Shepherd's-purse, and called it Life-Preserving Plant. It was credited with having the power to drive off mosquitoes and other nocturnal insect pests. Eating the plant was supposed to be beneficial to the liver and stomach, and at one time the poor ate large quantities of it as a potherb.

Suggested Use: Shepherd's-purse abounds in cities. It is a fine salad green, has a mild, pleasantly peppery taste, and is an excellent source of vitamins C and K. Chop the leaves coarsely and add them to salads.

SORREL *(Rumex acetosella)*

Folknames: Green Sauce, Common Sorrel, Field Sorrel, Sheep Sorrel, Sourgrass, Sourweed.

Location: Roadsides, thinly grassy areas, vacant lots, waste areas.

Botanical Description: Sorrel is rather an elegant little plant, generally between six and twelve inches tall. The jade-green leaves, arranged in a rosette when the plants are young, are arrow-shaped with winglike lobes at their bases. They are often tinged with red. The flower stems are slender, and several grow from the rather shallow roots. The flowers are extremely minute, grow in tiny clusters, and are quite beautiful when viewed through a hand lens. Male and female flowers are borne on separate plants. The plants are highly variable as to size, leaf shape, and color.

Sorrel is a perennial. The root system is extensive though shallow, and the plant reproduces itself by spreading underground runners as well as from seeds. The plants prefer an acid soil; their presence is an indicator that the soil is of poor quality. In such an environment, sorrel has very little competition from other plants, and thrives quite nicely.

Sorrel is an alien, introduced from Europe. It has become naturalized and grows throughout the United States. It is a member of the Buckwheat family and is related to rhubarb.

Historical Lore, Legends, and Uses: An old writer observed that Sorrel "may metaphorically be said to pierce the tongue like a razor with its sharpness." This is poetic exaggeration, but the scientific name does allude to this. *Rumex* is derived from the Latin *rumo,* "to suck," because the leaves were sucked to relieve thirst. *Acetosa* is derived from the Latin *acetum,* "vinegar," and refers to the acidic, vinegarlike taste due to the presence of oxalic acid in the leaves.

Sorrel has been a friend to man since ancient times. The old writers all praised its "grateful taste." In Tudor times, Sorrel was an important salad plant and pot-herb, and an esteemed medicinal plant as well. No herb garden was complete without its Sorrel patch, and it was one of the first seeds brought to the New World by the European settlers.

The leaves of Sorrel were known to be cooling and were therefore useful in fevers and a good gargle for soothing inflamed mucous membranes. The seeds drunk in wine and water were considered wholesome against the "frettings of the guts" and helped relieve a full stomach. They were also added to "binding medicines." The roots were used to treat scurvy.

"The juice of Sorrel in the summertime," remarked Dr. William Coles, "is a profitable sauce in many meats and pleasant to the taste, especially if some sugar is added. It cools a hot stomach, moves the appetite to food and refreshes the spirits being almost spent with the violence of furious, fiery fits of agues and quenches the thirst in them. For which there is nothing better," he advises, "than a Sorrel-posset drink: add the juice to milk when it begins to boil." Another soothing drink was prepared with Sorrel leaves: "A good quantity strained into ale and a posset made thereof cools and helps the thirst of those with pestilent fevers."

The juice of the entire plant mixed with a little vinegar was good to soothe itchy skin. Sorrel was considered a plague preventative. The leaves "eaten in the morning fasting, preserves one from Pestilence, in the time of it, and the leaves roasted in the embers and applied to the boil or plague sore cures it," we are told, "if it be applied in time."

The Anglo-Saxons of tenth-century England were familiar with Sorrel. They used it to treat infections. "Take this wort and old Swine lard, and the crumb of an oven-baked loaf. Pound together in the manner one makes a poultice. Lay it to the sore. It healeth wonderfully." Sorrel was also included in an Anglo-Saxon charm to cure a horse or other animal that was "elf-shot"—a name given to diseases caused by the machination of witches and other supernatural agents. "Take Sorrel seed and Scottish wax, let a man sing twelve Masses over it and put Holy Water on the animal. Have the worts always with thee."

Sorrel was enjoyed as a condiment too. Green Sauce, an old folkname, is a reminder of a favorite way of eating it in Tudor times. The fresh leaves were mashed with vinegar and sugar to the consistency of a sauce and served as an accompaniment for cold meats.

John Gerard mentions that the leaves of Sorrel were eaten in a tart, adding: "Sorrel is much used in sauces both for the whole and the sick. It is a pleasant relish for the well, quickening up a dull stomach that is overloaded with everyday's plenty of dishes. It is divers ways dressed by cooks to please their master's stomachs."

The Mohegans of Connecticut called the plant Sourweed. The fresh leaves were chewed and said to be good for the stomach. The Pennsylvania Dutch called Sorrel

SORREL: (*Below*) *young shoot,*
and (*right*) *entire plant with seeds,*
both approximately life size.

Sauerrampel. They made a diuretic tea from the leaves and ate leaves in salads to prevent scurvy.

In the Ozark Mountains, Sorrel has remained in use as a folk cure and is still used to treat a variety of skin diseases. A poultice of boiled, crushed leaves was freely applied to the afflicted areas and Vance Randolph says that it did cure many sores, but was very painful. (Sorrel contains significant amounts of oxalic acid, a powerful germicide. It may well have benefited minor skin troubles.)

In France, modern homeopathic physicians apply poultices of Sorrel leaves to bring boils to a head and help heal them.

Chinese herb doctors were quite familiar with Sorrel. They prescribed it to be taken internally for reducing fevers, and externally for treating ringworm (actually a fungus and not caused by worms) and other skin conditions.

Several varieties of Sorrel are cultivated here and abroad for use in soups, salads, and ragouts. In Lapland, Sorrel is used to sour milk. There was an old belief in New Brunswick, Canada, that eating "cow serl," as it was called, would give one head lice! This is entirely untrue and should not prevent anyone from eating this tasty herb. Sorrel grows abundantly in the city.

Suggested Uses: Sorrel can be gathered quite early in spring. It can be eaten throughout the year if you dig up a few plants in the fall (Sorrel likes acid soil, so bring along some of the soil they were growing in) and put them into flowerpots. Keep the soil moist and cut back flowering stems as they appear. This encourages the growth of new leaves, which can be picked and eaten as desired.

The quantities of Sorrel listed for the following recipes are safe to eat, but Sorrel should not be consumed in large amounts because the presence of oxalic acid in the leaves can irritate the kidneys.

SORREL TART WITH ONION AND BACON

This recipe serves six as an appetizer or serves four as a main dish.

1 small onion, finely chopped, 2 tablespoons butter, 4–6 slices bacon, 3 eggs (plus 1 yolk),
9-inch unbaked pie shell, 2 cups chopped Sorrel leaves, salt and pepper,
1 cup cream (light, heavy, or half-and-half), freshly grated Parmesan cheese (optional)

Sauté onion in butter until transparent. Set aside. Fry bacon until crisp, set aside Separate one of the eggs, and with a pastry brush paint the bottom and sides of the pie shell with the egg white. Let dry. (This prevents the filling from making the pastry soggy.) Mix eggs, adding extra yolk. Add Sorrel, onion, salt and pepper to taste, and cream. Pour into pastry shell and sprinkle with crumbled bacon and Parmesan cheese, if desired. Bake in a preheated 375° oven for about 35 minutes or until a knife inserted near the center comes out clean. Serve warm.

SORREL AND POTATO FRITATTA

This recipe serves six as an appetizer and four as a main course.

2 medium potatoes, vegetable oil for frying, 2 cups chopped Sorrel leaves, rosemary,
4 or 5 eggs, salt and freshly ground pepper, 3 scallions

Peel and slice potatoes. Fry in oil in a large ovenproof frying pan until brown on both sides. Add Sorrel and rosemary to the pan, and fry briefly. Beat the eggs, add salt and pepper to taste, and pour over Sorrel and potatoes. Cook until edges are set. Chop the scallions (both green and white parts), sprinkle them over the eggs, and slide pan under the broiler. Cook until top of fritatta is set, puffy, and lightly browned. Serve piping hot.

SORREL SOUP

Serves six to eight.

1 tablespoon butter, 1 small onion, finely chopped, 3 cups chopped Sorrel leaves,
5 cups chicken broth (do not use powdered concentrate) or water,
1 cup light cream or half-and-half, salt and freshly ground pepper

Melt butter, add onion, and sauté a few minutes until onion is golden. Add chopped Sorrel and sauté a few minutes. Bring broth or water to a boil, add it to Sorrel, and simmer very gently for about 15 minutes. Add cream, season to taste, and heat gently. Serve piping hot.

If desired, the cream can be omitted entirely. In this case, it is nice to add a well-mashed boiled potato to the soup as it heats, to enrich and thicken it.

SORREL SKIN WASH

The juice of Sorrel leaves steeped in water is reputed to make the skin smooth and clear.

Chop 1 cupful of fresh, washed Sorrel leaves. Put them in a bowl and pour 3 cups of boiling water over them. Cool to tepid, then strain. This wash should be patted gently over the face and neck. It can also be added to the bath.

SOW THISTLE *(Sonchus species)*

Folknames: Hare's Thistle, Hare's Lettuce.

Location: Roadsides, vacant lots, waste areas, throughout.

Botanical Description: Of the six species of Sow Thistle that grow in the United States, three varieties are seen with particular frequency in the city.

Common Sow Thistle (*Sonchus oleraceus*) grows from one to three feet tall. Its stems are hollow, branched, and rather thick. The leaves clasp the stems and are oblong, deeply lobed, and have irregular, somewhat prickly teeth along the edges. The upper leaves are much smaller and simpler in shape.

Field Sow Thistle (*Sonchus arvensis*) looks very much like Common Sow Thistle, except that it has hairs on the stalks and prickly leaves.

Prickly Sow Thistle (*Sonchus asper*) has sharply curved leaves with earlobelike bases that clasp the stems. The leaf edges are very spiny and often curled or wavy.

All three species of Sow Thistle have small yellow flowers that resemble those of the Dandelion, to which they are related. The seeds are also similar, being attached in the same manner as Dandelion to fluffy white "parachutes," easily dispersed by the wind. All three Sow Thistles have sticky white sap that flows freely when any part of the plant is broken.

Prickly and Common Sow Thistles are annuals and reproduce by seeds. Field Sow Thistle, a perennial, reproduces by seeds and runners growing horizontally underground, close to the soil surface. These break off easily, and each piece can produce a new plant. The perennial form is very difficult to eradicate and particularly disliked by farmers.

All the Sow Thistles are aliens and were introduced from Europe. They have become naturalized and are common throughout the United States. Sow Thistles are generally dismissed as "coarse, unattractive weeds," and the perennial variety is invariably referred to as "noxious." One's attitude to them in urban environments ought to be more generous, however. Anything that flowers in otherwise neglected waste areas in the city should be more than welcome.

Sow Thistle plants are actually quite fascinating if one takes the time to examine them closely. The leaves of the Prickly Sow Thistle are particularly remarkable, even beautiful. The graceful curve and shape of an individual leaf, its complex vein pattern, the arrangement and structure of the prickly teeth along the edge of the leaf, and the manner in which it clasps the stalk are fascinating when carefully observed. There is truly a world of nature waiting to be discovered in a single leaf of this plant.

Historical Lore, Legends, and Uses: Common, Field, and Prickly Sow Thistle closely resemble one another, and all three greatly resemble Wild Lettuce, which has several varieties of its own. This led to some confusion in the centuries prior to the Linnean system of classification. The seventeenth-century herbalist Dr. William Coles grouped them all together and believed they had similar if not almost identical virtues. In point of fact, Sow Thistles and Wild Lettuces are related; both are members of the *Compositae* family. Wild lettuces are members of the genus *Lactua*, however, and Sow Thistles are members of the genus *Sonchus*.

The manner in which Sow Thistle received its common name is rather charmingly related by Dr. Coles: "When sows have pigs," he explains, "they greedily desire it, because they know by certain natural instinct that it very much increases their milk. For that reason, I conceived it is called Sow Thistle." The generic name *Sonchus* is derived from an ancient Greek word meaning "hollow," alluding to the structure of the stems.

The Greeks and Romans knew Sow Thistles and considered them wholesome and beneficial additions to the diet and valuable medicinally. The milky sap was recommended for eye ailments and skin eruptions, and was dissolved in drinks for "shortwindedness" and wheezing. The distilled water of the plant made the complexion soft and clear. Taken with a little sugar, this liquid was a medicine "the daintiest of stomachs" would not refuse. The fresh leaves were also chewed to sweeten offensive breath.

The Anglo-Saxons ate Sow Thistles in broth to heal "gnawings of the stomach." To expedite childbirth, an expectant mother was fed three spoonfuls of the juice of Sow Thistles dissolved in warm white wine. The Anglo-Saxons believed that eating Sow Thistles "caused milk in nurses and a good color in their children," that is, in the children they nursed.

Seven hundred years later, seventeenth-century herbalists who adhered to the doctrine of signatures claimed that a decoction of the leaves and stalks of Sow Thistle caused "an abundance of milk" in nurses and was "good for those whose milk curdles in the breast." According to the theory, the white milklike juice of Sow Thistles indicated that the plant would benefit nursing women.

Sow Thistles have been greatly appreciated as a potherb since very early times, though to modern tastes they would probably seem quite bitter. Common Sow Thistle was probably eaten, since it had the least prickles and was more succulent than the others. Theseus is supposed to have been fed a dish of Sow Thistles to

SOW THISTLE:
(*Left*) *leafy stalk,*
and (*above*) *flowering stalk*
from upper plant,
both approximately life size.

strengthen him before his encounter with the Bull of Marathon. It was said that Sow Thistles were "very fit to cool a hot stomach and ease gnawing pains thereof, and are therefore eaten by some as a sallet herb in winter and spring."

Parkinson noted that the young leaves of Sow Thistle, especially the smooth kind, were less bitter than the prickly varieties and were eaten in the spring as a "sallet." Sow Thistles were perhaps better known as a potherb on the Continent. The Italians were said to relish the roots and leaves of the young plants in salad, and Dr. William Coles mentions that the plants "when young and tender were eaten familiarly by those beyond the sea."

Sow Thistle, perhaps because of its rather hurtful, prickly spines, has been a plant long associated with sorcery and magic. The Russians believed it to be the devil's favorite plant, and it was an herb of Venus, making it useful in love matters.

Sow Thistle was reputed to reveal treasures in a manner similar to the sesame of Oriental folklore. There are Italian legends in which the invocation "Open, Sow Thistle!" give the very same result.

Sow Thistles were well known to the Chinese. They were classed with a group of plants referred to as "bitter vegetables." Chinese herbalists credited Sow Thistle with powerful medicinal properties. The juice of the plant was used to treat abscesses, boils, and warts. The yellow flowers were considered to have calming power, and remedies to reduce fevers were prepared from them. Prolonged use of Sow Thistle was beneficial in conserving health and vitality.

The use of Sow Thistle as a medicinal plant in Europe was eventually abandoned. By the middle of the nineteenth century, Dr. Thomas Green noted that it was little regarded as a medicine, even though it was recognized to have nearly the same properties as Dandelion and Chicory, both useful medicinal plants in his time. Dr. Green remarked that Sow Thistles were a great favorite with rabbits, were eaten by goats, sheep, and hogs, but that horses disliked them.

Suggested Uses: Sow Thistles should be gathered for eating early in the spring, before they become too bitter. The young raw leaves can be added to salads.

As a potherb, cook the young leaves for 5 minutes in boiling water to cover. Drain and add fresh boiling water to cover, and a little salt. Cook 5 minutes more, drain, and serve with melted butter.

Prickly Sow Thistle leaf.

SWEET MELILOT
(Melilotus alba, M. officinalis)

Folknames: Sweetclover, King's Clover, Plaster Clover, Sweet Lucerne, Hart's Clover.

Location: Roadsides, vacant lots, waste areas.

Botanical Description: Sweet Melilot is very common in the city. It is rather tall, usually growing between three and five feet high. The plants are somewhat bushlike, with graceful, many-branched stems. The leaves are compound and have three cloverlike, elongated leaflets, finely toothed at the edges.

Melilot blooms from June to October in the northeastern part of the country. The flowers are borne on rather long, curved stems. The individual flowers are quite small, and the beauty and delicacy of their structure, similar to Clover flowers, can best be observed with the aid of a hand lens. Two varieties of Melilot are common in cities, and they often grow side by side. *Melilotus officinalis* has yellow flowers; *Melilotus alba* has white flowers. Both are biennials and reproduce from seeds capable of remaining dormant in the soil for many years.

Sweet Melilot is fragrant when fresh, but the aroma of the foliage is greatly intensified when the plants are dry. This is due to the presence of coumarin, a substance that imparts to Melilot its characteristic aroma of "new-mown hay."

Despite the common name Sweetclover, Melilots are members of the genus *Melilotus* and are not actually true Clovers (*Trifolium*). Melilots and Clovers are both members of the pea family (*Leguminosae*), however, and have nitrogen-fixing bacteria that grow in association with their root systems. These bacteria enrich the soil and make Melilots valuable agents in the reclamation of the generally poor soils they thrive in.

Melilot is an introduced alien. It has become naturalized in the United States and now grows everywhere.

Historical Lore, Legends, and Uses: As a medicinal, forage, and bee plant, Sweet Melilot has been known since antiquity. The ancient Egyptians are known to have used preparations of the plant for treating intestinal worms and earache. *Mel* means "honey" in Greek, and the name Melilot is derived from the Greek word for it: *Mellita lotus,* a "lotus that smelled like honey." Bees love this plant, and it is an important nectar source for them.

The classical origins of several of the other names for Sweet Melilot are equally picturesque. An old Latin name for Melilot was *Trifolium equinum,* or Horse's Trefoil, because horses loved it and grew fat eating the plants. In later centuries Melilot was a valuable forage plant, especially popular in Tudor times. It was variously called Hart's Clover, because stags "delight to feed upon it," and King's Clover, because it was "chiefest of all the three-leaved grasses." John Parkinson commented that in Essex, Melilot was named Heartwort: "The folk there think the seed thereof happening into their bread causes pains in the stomach and chest which they usually call the heartburn."

Parkinson wrote that insofar as medicinal virtues of Melilot were concerned, "Arabian physicians do use the seeds and seedpods and do not mention the flowers: the Greeks, contrariwise, do will the flowers only to be used and never use the seeds or pods." During the first century A.D. Galen prescribed a poultice containing Melilot which he recommended for inflammations and swollen joints.

This prescription may have been the origin of a salve that for centuries was called simply Melilot. It was compounded of the juice of young green Melilot plants boiled with rosin, wax, sheep tallow, and a little turpentine. Herbalists claimed that if well made, it would be almost as green as the herb itself and smell very strong, "though it be two or three years old." This salve or plaster was used to draw and heal all kinds of wounds and sores and remained popular for centuries. A similar Melilot plaster can still be purchased today in many parts of Europe.

A "most sovereign balm" for sore muscles and general aches and pains was prepared by adding the fresh leaves and flowers of Melilot to olive oil. The mixture was then steeped in the sun and strained, and fresh plant material was added. This process was repeated, and finally the resultant oil was boiled and bottled. Melilot was said by the old herbalists to have the power to "bind together" and soften all inflammations. Sometimes it was mixed with the yolk of a roasted egg, or linseed, Chicory, or poppy heads.

The juice of Melilot was considered "good to strengthen the head and brains." Dr. William Coles advised that it was effective "applied to those who have suddenly lost their senses, by any paroxysm, also to strengthen the memory, comfort the head and brains and preserve them from pain and fear of apoplexy, if the head be washed often with the distilled water of the herb and flowers."

The Anglo-Saxons believed Melilot preserved the eyesight, and a charm whereby this could be accomplished is recorded in a tenth-century leechbook: "This wort thou shall take up in the waning of the moon, in the month of August. Take then the root of this wort and bind it to a yarn thread and hang it to thy neck; that year thou shalt not feel the dimness of thine eyes, or if it befall thee, it shall suddenly

SWEET MELILOT: *Flowering stem,*
approximately life size.

depart, and thou shalt be whole. This Leechcraft," they add, "is a proved one."

Corn Melilot (*Melilotus arvenis*), similar and closely related to Sweet Melilot, is common in China, especially in the Yangtze provinces. It used to be burned in ancient China because it was believed to have the power of summoning ancestral spirits. Worn on the person, the plant dispelled unwholesome influences. Corn Melilot was a useful medicinal plant as well. It was considered sedative and astringent. When taken internally, it imparted its sweet fragrance to the body.

Seen everywhere in the city and at home in the most inhospitable environments, Sweet Melilot was the star in a drama that led to the discovery of an anticoagulant whose importance in modern medicine cannot be overestimated. The story is a remarkable one and exemplifies scientific detective work at its very best.

It has long been known that Melilot, or Sweetclover, as farmers call it, makes fine hay and that cows who eat it produce rich milk, butter, and cheese. A popular forage crop in Europe, Sweet Melilot came to the attention of agriculturists in the United States during the 1920s and soon became a widely employed soil builder and forage crop.

Sometime after its introduction as feed, farmers noticed their cattle were developing a strange disease. They began to hemorrhage internally and externally, and the bleeding was so severe that sometimes the creatures died. The problem was serious and alarming, and scientists immediately set to work on it. After a great deal of research, it was determined that the culprit was Sweet Melilot.

Melilot hay was both harmless and nutritious when stored after it had been dried. But if it was stored for later feed with a moisture content of about 50 percent, it readily became moldy and toxic as well. At first, it was thought the mold itself was responsible for the strange bleeding disease, but further research and testing showed this was not the case. Instead, it was discovered that both white- and yellow-flowering species of Melilot contain a chemical compound derived from coumarin. When the plants were stored moist, and as a result spoiled, the normally harmless coumarin underwent a chemical change and became dicoumarol, a substance that reduces blood prothrombin, a vital clotting agent. If large amounts of spoiled Melilot were eaten, enough dicoumarol was ingested to render the blood incapable of coagulating, and death often resulted from uncontrolled hemorrhaging.

After a great deal of work, scientists developed a procedure whereby the isolation and identification of dicoumarol was accomplished. This substance is now manufactured synthetically and is one of our most important blood-thinning drugs. Dosage must be very carefully regulated, however. The formation of dangerous, possibly lethal, blood clots must be prevented, yet the blood cannot be so thinned out that serious hemorrhage occurs.

Curiously, the fundamental reaction whereby this substance prevents coagulation of the blood is not fully understood. It *is* known, however, that vitamin K acts in virtual opposition to it, enabling this vitamin to be of great service in counteracting overdoses. In addition to its medical value, this chemical is a highly effective ingredient in warfarin, one of the most powerful modern rat poisons. Here it is particularly useful, since rats have been unable to develop an immunity to it, as they are easily able to do with other materials. This problem made difficult the development of effective means of exterminating these baneful and disease-carrying animals. Warfarin was therefore an important weapon in this continual battle.

For the sake of its sweet fragrance, Melilot has long been employed to perfume snuff and tobacco mixtures. At one time the distilled water of the flowering plants was added to the rinse water for clothes and linens to make them smell sweet, and sprigs of dried Melilot were laid away with furs to give them a pleasant fragrance and keep the moths away.

Suggested Uses

MELILOT CHEESE

The Swiss prepare a cheese called *Schabzeiger* (Sapsago), which is flavored with a different but closely related variety of Melilot than grows here. The Swiss call this plant curd herb. It is dried, powdered, and added to the cheese, which is then cured in special molds until it becomes quite hard. When wanted for use, the cheese is grated, mixed with fresh butter, and spread on bread. You might wish to sample my adaptation:

Allow an 8-ounce package of cream cheese to reach room temperature. Blend in about 1 tablespoon (or a bit more—taste it first) of dried, powdered Melilot leaves. Mix well and let mellow several hours or overnight in the refrigerator. Serve with whole-grain bread.

MELILOT SKIN WASH

Pour 2 cups of boiling water over about 1 cup of fresh Melilot flowers. Cover and let cool until tepid. Strain, pressing the flowers to extract as much liquid as possible. Pat gently on the skin. This wash is soothing and helps heal chapped or inflamed skin. It can be added to the bath. Gauze pads wrung out in it are good compresses for tired eyes.

MELILOT CHEESE

Pour 1 cup of boiling water over 2 tablespoons of the dried flowering tips of Melilot; use either the yellow variety or the white. Let steep, covered, about 10 minutes. Strain and sweeten with honey. Lemon may be added. This beverage is very relaxing and quite pleasant tasting. It is especially good if taken before bedtime.

To dry Sweet Melilot: Gather the plants anytime they are in bloom and hang up to dry in a shady place where there is good air circulation. When completely dry to the touch, strip leaves and flowers from the stems and store in jars away from the light until wanted for use.

Dried Sweet Melilot leaves and flowers are excellent additions to homemade potpourri and sachet mixtures.

AN ELIZABETHAN "QUILT BAG"

I have not tried this old Elizabethan remedy for "easing pain in the sides," but it would seem from the list of ingredients that, in any event, the aroma would be comforting.

"Take Melilot flowers, Chamomile, Rosemary, Elder flowers, of each a handful. Bran a handful, of anise seed, fennel seed, caraway seed, bruised, 2 ounces each. Have a quilted bag for the purpose, fill it, sprinkle it with wine and being made hot, apply it."

WILD SUMAC *(Rhus glabra)*

Folknames: Smooth Sumac, Lemonade Sumac, Pennsylvania Sumac.

Location: Vacant lots, road embankments, railroad rights-of-way, parks.

Botanical Description: Wild Sumac is a sparsely branched shrub or small tree. In urban environments it may grow only between four and six feet high, but it can be eight or more feet tall in favorable situations. It has compound leaves with eleven to thirty-one leaflets borne on a single long, red leaf stalk. These leaflets are small, stemless, and have serrated edges. They are pointed at the tip, and the undersides have a whitish bloom.

In the Northeast, Sumac blossoms in July. The flowers are in thickly clustered masses at the ends of the stems and are fragrant. The individual flowers are rather small and greenish-white to yellow in color. After the flowers are gone, they are followed by dense, cone-shaped clusters of red berries, conspicuous and quite attractive during the summer and fall. The velvety appearance of the berries is due to the presence of hundreds of tiny, sticky red hairs, with which they are covered. When ripe, these berries emit a fragrant, delicious aroma of ripe raspberries.

Two additional varieties of Sumac are sometimes encountered growing in the same environment as Wild Sumac. All three are quite closely related and resemble each other.

Staghorn or Velvet Sumac (*Rhus typhina*) looks very much like Wild Sumac. It has velvety red stems, whereas the stems of Wild Sumac are smooth.

Dwarf Sumac (*Rhus copallina*) is somewhat smaller than Wild or Staghorn Sumac. The leaves are composed of nine to eleven leaflets with flattened, winglike membranes on either side of the leaf stalk between the individual leaflets. The fruit clusters are somewhat drooping, rather than erect as in the other two species.

These three varieties of Sumac are native. They are perennials and reproduce by seeds. There are several other varieties of Sumac native to southern Europe, the Mediterranean countries, China, and Japan. All species exude a sticky, milklike sap if the stems or leaves are broken. The foliage of all the Sumacs turns beautiful shades of crimson and red-orange in the autumn, making them especially welcome in the city. After the leaves fall, the red berry clusters persist, often throughout the winter.

Wild, Dwarf, and Staghorn Sumacs are perfectly harmless, yet most people shun these pretty and useful plants entirely, believing them to be poisonous. The much-feared Poison Sumac (*Rhus vernix*) is said to produce a worse rash than Poison Ivy (to which it is very closely related), but it grows in wet, swampy places and is seldom, if ever, encountered in the city.

In any event, Poison Sumac is easily distinguished from the other two. It has *greenish, cream-colored, or white berries borne in drooping, loosely clustered groups* at the joint of the leaf and stem. The compound leaves have only *nine to eleven leaflets* on the leaf stalk. Poison Sumac is indigenous to the United States, and the Iroquois used it medicinally.

Historical Lore, Legends, and Uses: Sumac contains tannic, citric, and malic acids and is therefore quite astringent. The plants have been used as a medicinal herb and dye source for centuries. Sumacs were known to the ancient Greeks and other classical scholars, and their medicinal properties were highly esteemed. During the first century B.C., the physician Galen (whose word was law for fifteen centuries) recommended a broth of Sumac to control bloody fluxes and poultices of the fresh leaves to treat gangrene. In Galen's time, Roman women used a strong infusion of the leaves to dye their hair black.

William Coles described several species of Sumac, used as medicine in his day (seventeenth-century England). He noted that "the reddish color of the seeds did teach those that found the virtues of plants by their signatures, that it is good for the bloody flux. Upon this account," he remarked sarcastically, "it is prescribed gen-

WILD SUMAC: *Stem with berry spike,
slightly less than life size.*

erally by those who meddle with the cures of diseases." He added that Sumac leaves were boiled in wine and drunk, and the seeds were eaten in sauces or with meat.

A decoction of Sumac was said to be good to set in a convenient place "where the fumes may powerfully ascend into the bodies of those diseased." The seeds, crushed and boiled into a poultice, healed inflammations and were good to make into a wash for the head in cases of "want of sleep." A decoction of the seeds was excellent for the mouth and throat. It healed sore gums and was said to even fasten loose teeth! The juice of the dried leaves and seeds boiled in water and then evaporated to the "thickness of honey" also helped sore mouth and throat and was believed even more effective than preparations of the seeds alone. The juice of the fresh green leaves dried up running ears.

Nineteenth-century American physicians frequently prescribed preparations made from Sumac. The berries were considered "powerfully styptic" and helped stop bleeding. The bark was recommended for treating skin irritations. For this purpose an ointment was made by simmering the bark of the roots with lard. Dr. C. S. Rafinesque, a nineteenth-century American physician and botanist, observed that this salve healed burns without leaving scars. This doctor reported that dried Sumac leaves made an excellent substitute for tobacco, and that smoking them caused those who did so to loathe tobacco afterward. Spirituous infusions of Sumac were rubbed on the limbs to relieve rheumatism and aching muscles, and small balls of the gummy sap inserted into tooth cavities relieved the pain of toothache.

Until very recently, Sumac was used medicinally in the United States. A drug derived from the dried ripe berries was used as an ingredient in gargles. The 1917 edition of *Potter's Therapeutics* lists "Fluidextract of Rhus Glabra" as a "useful astringent gargle" and adds that an infusion of the fluid extract with water could be used as a wash for ulcers and wounds.

Native American peoples were very familiar with Wild Sumac. Many parts of the plant were used to prepare medicines for a variety of ailments. The Delaware used the entire plant (with the exception of the leaves) to cure sore gums and canker sores in the mouth. The roots were boiled and used as poultices for toothaches. The Potawatomi Indians used the leaves to make a gargle for sore throats and prepared a medicinal tea from the berries.

Other tribes made a cooling drink by infusing the ripe berries in water, while the hairs on the berries were used as a salt substitute. A medicine called *kinekah,* used by several western tribes, contained powdered leaves and roots of Sumac. This was mixed with tobacco and used to treat a number of conditions.

Sweet Sumac (*Rhus aromatica*) is a western species with fragrant wood. The native Americans of New Mexico, California, and Arizona used it for basketmaking.

Other tribes obtained red dye from the roots of the Sumacs and made flutes from the stems by pushing out the soft woody pith with a thin stick.

In the Ozarks, Sumac leaves were reputed to cure hayfever and asthma; they were dried and smoked in a pipe or brewed into a tea, and tea was made from the berries as well. This tea was drunk to ease colds and sore throats and was supposed to cure children of bedwetting.

Various European species of Sumac have been used for centuries in the production of fine leather goods and textiles. The leaves of the Mediterranean varieties have a great deal of tannin, and black, red, and yellow dyes were obtained from them. Plantations of Sumac grew in Italy, Spain, and Syria. William Coles says that the dried leaves were sold for these purposes in markets, where they fetched "huge sums of money."

In the United States as well as in Europe, Sumac at one time was an extremely valuable source of a very fine black dye for silk and wool. This may not seem very important today, but it should be remembered that black was the most difficult color of all to obtain from plant materials. It was not until the middle of the nineteenth century that the cheaper, more reliable, and simpler-to-use aniline dyes became widely available. These chemical dyes quickly replaced the vegetable dyes that had been in use for thousands of years.

All the Sumacs are members of the family *Anacardiaceae*. They are closely related to Poison Ivy and are members of the same genus, *Rhus*. Several other members of this genus are indigenous to the Orient and grow there in abundance. One of these is the source of the varnish used to make the beautiful and world-renowned Japanese lacquerware. The sap of this plant, known as the Japanese lacquer tree, is collected from cuts made in the bark of three-year-old trees. When fresh, the sap is reputed to cause a severe Poison Ivy–like rash if it comes in contact with the skin, and great care is taken in the gathering of it.

At first, the sap is light-colored, with the consistency of heavy cream. It becomes thicker upon exposure to air. A varnish prepared from this sap is so transparent that, laid pure and unmixed on wood, it brings out the finest and most delicate features of the grain. For the manufacture of lacquerware, a dark ground is applied first. This undercoat (usually red or black, but sometimes gold leaf is used) causes the lacquer to reflect the light like a fine mirror. When dry, the lacquer becomes extremely hard but has peculiar properties. It cannot sustain direct blows, which will cause it to shatter like glass, but boiling water will not harm it in the least.

Maude Grieve, a modern herbalist, mentions a material called oil of *Rhus* that can be extracted from the seeds of the American Wild Sumac and made into candles. She remarks that they burn brightly, but have a very pungent smoke.

WILD SUMAC DRINK (SUMAC LEMONADE)

Sumac berries are a fine source of vitamin C, and an excellent, very refreshing beverage with a delicious thirst-quenching taste (even better than lemonade, in my opinion!) can be easily prepared from the ripe berries (free for the picking) of Staghorn or Wild Sumac plants. Gather about ten or fifteen ripe berry clusters. Discard any leaves, rinse the berry heads under cool water, and strip them from their stems into a bowl.

Pour about 2 quarts of water into a large bowl. Now take a handful of berries at a time and rub briskly between your hands, letting the berries fall into the bowl of water as you do so. (This loosens the tiny hairs covering the berries and releases their flavor.) Do this until all the berries are used up. Do not be alarmed if your hands become red—it will wash off. Strain the liquid (it will be a lovely shade of rose-pink) through coffee filter paper to remove all the fine hairs. Sweeten the beverage to taste with honey or sugar and serve well chilled.

Wild Sumac is a fine source of natural dye. Shades of tan and brown can be obtained from the leaves and young shoots. See the Appendix, page 172, for instructions on preparing dye.

YARROW *(Achillea millefolium)*

Folknames: Yarroway, Milfoil, Thousandweed, Thousandleaf, Soldier's Woundwort, Knight's Milfoil, Carpenter's Weed, Bloodwort, Staunchweed, Sanguinary, Nosebleed, Devil's Nettle, Devil's Plaything.

Location: Roadsides, grassy strips, waste places, vacant lots.

Botanical Description: Yarrow is a plant whose growth habits are rather variable. One or several stiff stems may grow from the root. They are usually between one and two feet tall but are occasionally shorter, and may be smooth- or rough-textured. The leaves are larger at the base and progressively smaller toward the top of the stalks and are arranged alternately. They clasp the stems at their bases and are delicate and finely divided, resembling feathers more than leaves.

Yarrow blooms from June to September in the eastern part of the United States. Flowers are in flat-topped clusters at the ends of the stems. The individual "flowers" are very small, with fine white "petals" and a yellowish center. This "flower" is actually two separate, distinct male and female flowers. The female flowers are in the yellow center surrounded by five white "petals," each one of which is a male flower. The entire plant has a strong, pungent odor and a bitter taste. If Yarrow is eaten by cows (as occasionally happens) it gives a very unpleasant taste to milk products and makes them inedible.

Yarrow is a perennial, reproducing by seeds and from underground runners. It is native and widespread in the United States and grows throughout Europe and Asia as well.

There is a variety of Yarrow that has beautiful purple flowers. It is grown in gardens and is called Ornamental Yarrow. This color is sometimes (but rarely) found in wild plants as well.

Historical Lore, Legends, and Uses: Legend has it that Achilles was taught the medicinal virtues of Yarrow by the centaur Charon, who was skilled in herb lore. Achilles used the herb to heal his soldiers' bleeding wounds, and the herb was named in his honor, though some might feel it should have been named after Charon. In any event, Yarrow has an ancient and honorable reputation as a wound herb, particularly efficacious for stopping the flow of blood. This belief was alluded to in the folknames, most of which refer to this property. The specific name, *millefolium,* refers to the minutely divided leaves and is reflected in names such as Milfoil and Thousandleaf, a literal translation of the Latin word *millefolium.*

Yarrow was said to be "excellent to stop inward bleeding." Yarrow was dried, powdered, and mixed with Plantain or comfrey water (both were famous wound herbs) or used by itself fresh, as a poultice for wounds that would not stop bleeding. These preparations were said to immediately stop the flow of blood. Dried and powdered Yarrow leaves, if dropped into the nostrils, stopped nosebleed. A decoction of Yarrow in white wine was drunk as a remedy for too copious menstruation. For the same purpose, large amounts of the fresh plants were boiled in water, and the patient sat over the beneficial steam to absorb it.

Oddly enough, this stauncher of blood could actually *cause* nosebleed if a fresh leaf was inserted into the nostril and twisted. This was sometimes purposely done, it being believed at one time that nosebleeds cured headaches.

Yarrow was a favorite wound herb of the Anglo-Saxons. They also employed it to heal burns and the bites of poisonous snakes and insects. The fresh leaves were chewed to relieve toothache.

Dr. William Coles, a seventeenth-century physician, prescribed the flowers and juice of the plant taken in goat's milk or the distilled water of the whole plant as being "good for loose bowels, even more so if a little powdered coral, amber or ivory is added." (This last recommendation was medically worthless but highly popular in Coles's time, particularly among the rich—the only ones who could afford it. Everyone else had to be content with the unadorned herbs. Actually, they were probably better off. Powdered gems certainly did no good, and in some instances may have done some harm.)

Coles mentions that ointments containing Yarrow were used to heal ulcers, wounds, and running sores "by signature—the many incisions upon the leaves resembling those wounds, or if your fancy will have it, more like unto hair: it stops the shedding if the head is bathed with a decoction thereof." He describes another more unusual use for Yarrow: the juice was injected by syringe to cure a distressing ailment known as "the excoriation of the yard [penis] caused by pollution or extreme flowing of seed, and any inflammation or swelling caused thereby, as has been proved by some single or unmarried persons, who have been very much oppressed on this account."

In nineteenth-century Britain, one physician observed that Yarrow "though generally neglected" was a fine medicine for excessive menstrual bleeding, bloody fluxes generally, and bleeding piles. It was an excellent diuretic and healed ulcers of the kidneys and urethra. The best part was the young shoots. The doctor remarked that foreign physicians still esteemed Yarrow for treating hemorrhage.

In America, Yarrow was well known as a medicinal plant to native American peoples. The Delaware and related Algonquin tribes prepared a tea from Yarrow which they used for treating liver and kidney disorders. The Lenape pounded Yarrow roots with a stone and boiled them with water to make a remedy for excessive menstrual flow. Yarrow was extensively employed by a number of other tribes. The Ute name for it meant "wound medicine," and it was used by them as such, and the Piute drank Yarrow tea to cure a variety of stomach disorders.

The Pennsylvania Dutch knew Yarrow as *Schoof Ribba*. They prepared a "sweating tonic" from the whole plant to reduce fever, and a tea made with the leaves was supposed to have a beneficial effect on the liver and gall bladder. Horses were fed Yarrow to cure them of intestinal worms.

The nineteenth-century physician-botanist Dr. C. S. Rafinesque recommended an infusion or extract of the whole herb for menstrual problems and dysentery. Rafinesque believed that American Yarrow was stronger in its action than the European variety, and he mentioned that the American plants were exported for medicinal use abroad.

YARROW: *Flowering stalk,*
approximately life size.

Yarrow tea was a popular remedy for influenza. It was thought to induce copious sweating, thus reducing the dangerously high fever of this disease. Yarrow tea was considered a good general remedy for severe chest colds as well. The dose was one ounce of the dried herb to one pint of boiling water. It was strained and drunk warm and sweetened with honey or sugar. Sometimes a dash of cayenne pepper was added.

Yarrow is still official in Central Europe as a tonic and stimulant.

The *British Herbal Pharmacopoeia* (1971 edition) lists Yarrow as an "antipyretic, diaphoretic . . . astringent and diuretic." Modern herbal doctors employ it to treat fevers, amenorrhea, and diarrhea.

Perhaps because of its pungent (and to many, unpleasant) odor, Yarrow was said to be one of the devil's herbs and was probably called Devil's Plaything and Devil's Nettle for this reason. In any event, it has been long associated with magic and witchcraft. As is so often the case, however, the plant could actually be employed to give protection against the very same spells that it was an ingredient of.

Yarrow was strewn across the threshold of a house to keep out evil influences and was worn to guard against evil spells. Country people tied sprigs of it to a baby's cradle to protect the infant from witches who might try to steal away its soul, which they believed to be a real possibility in cases where there had been a delay in baptizing the infant.

To ease childbirth, Yarrow that had been gathered on St. John's or Midsummer Eve (June 21, the summer solstice, a day of great and powerful magical significance since very ancient antiquity) was given to a woman in labor. She held it pressed to her right side, but it had to be taken away as soon as the child was born.

A strange Anglo-Saxon charm "for a fiend-sick man or demonic, when a devil possesses the man or controls him from within with disease" is recorded in one of the tenth-century leechbooks. The charm proceeds to describe the thirteen herbs needed, one of which was Yarrow, to be made into a "spew drink" (to cause vomiting—that is, the "vomiting out" of the evil) to be drunk from a church bell. Seven Masses were then sung over it, and garlic (an ancient protector against evil spirits) and holy water were added. Not only was this mixture to be drunk from a church bell—and one wonders exactly how this was accomplished—but the brew was to be added to everything the sick man ate or drank. Psalms 119, 67, and 69 were sung over it, it was drunk out of the church bell, and the Mass priest afterward said a benediction over the sick man. This was a complicated ritual, and from it one may infer that demonic possession was believed a reality and received serious attention. One wonders how frequently this procedure, with its peculiar combination of pagan and Christian elements, was resorted to.

Yarrow was a plant of Venus (this was odd, because most devil's herbs were plants of Saturn) and, as such, was frequently consulted where love matters were concerned. One famous love charm required that a handful of Yarrow be sewn into a flannel square and put under the pillow, and the following rhyme said aloud:

Thou pretty herb of Venus' tree,
Thy true name is Yarrow.
Now who my bosom friend must be,
Pray tell thou me tomorrow.

One's future husband or wife would appear that night in a dream.

Another love divination was based on Yarrow's well-known ability to cause nosebleed. A Yarrow leaf was inserted in the nostril and gently rotated while the following was recited:

Yarroway, Yarroway, bear a white blow [flower].
If my love loves me, my nose will bleed now.
If my love do not love me, it won't bleed a drop,
If my love loves me, 'twill bleed every drop.

Another charm recited to ensure the appearance of a future husband or lover in a dream was common in the south of England. A girl picked a sprig of Yarrow from the grave of a man who had died young, reciting:

Yarrow, sweet Yarrow, the first I have found,
In the name of Jesus Christ, I pluck it from the ground.
As Jesus loved sweet Mary and took her for his dear,
So in a dream this night,
I hope my true love will appear.

She then took the plant home and put it under her pillow. This charm is an odd one, to say the least. It might even be considered blasphemous. After all, Jesus did not take anyone "for his dear," much less someone named "sweet Mary"! The Virgin and numerous saints were frequently begged to intercede for or otherwise come to the aid of lovers, but as far as I know, few love charms invoke the name of Jesus. I do not know the history of this particular invocation, or how ancient it is, but a possible explanation is that the charm is indeed a very old one, and the names of Jesus and Mary were substituted (as was the case with many other charms) at a later date, to replace the names of pagan gods and goddesses.

Yarrow was frequently included in wedding bouquets and garlands, where its presence was said to guarantee true love between the married pair for seven years!

There were other beliefs associated with Yarrow. The juice, if rubbed into the hair, made it curly. To dream of it after gathering the plant for medicine meant the dreamer would hear good news. In the Orkney Islands of Scotland, Yarrow tea was a cure for melancholy, while in the Hebrides, a leaf of Yarrow held against the eyes gave "second sight."

Yarrow was considered a beneficial medicinal herb among the Chinese. It was said to be useful in improving respiration, skin, and muscle tone and if taken for a long while was believed to increase intelligence. It is called *shih* in Chinese and is said to grow in exceptionally plentiful amounts at the grave of Confucius. According to a Chinese legend, one hundred Yarrow stalks grew from a single root. When the plant was a thousand years old, three hundred stalks would grow from the root. Such was the power of this plant that wolves, tigers, and poisonous plants would never be found near it.

Yarrow has special significance to the consulters of the *I Ching*. Stalks from a closely related species (*Achillea sibirica*) are the source of the famous "stalks of divination" to be used in consulting this oracle. These stalks were sold in parcels of sixty-four, and their length was very important. For the Son of Heaven (the emperor) the stalks were nine feet long; for feudal princes, seven feet; for high dignitaries and government officials, five feet long; and for graduates (probably of the mandarinate), three feet.

Suggested Uses: Modern herbal doctors employ preparations of Yarrow to treat pneumonia, nephritis, and Bright's disease.

Yarrow is a good companion plant in the vegetable garden. Its root secretions are said to be strengthening to other plants and actually makes them more disease-resistant. Yarrow is also said to keep ants and harmful insects away.

YARROW SKIN WASH

Yarrow makes an excellent skin wash, its astringency making it particularly beneficial to oily complexions. Pour 2 cups of boiling water over about 1 cup of crumbled, dried flowering Yarrow tops, cool, and strain. Pat on the skin. This wash soothes chapping and minor irritations as well.

To dry Yarrow flowers: Gather freshly opened flowering stalks, breaking them off at the base. Tie in bunches of three or four and hang upside down to dry, in an airy place, away from direct sunlight. When they are thoroughly dry, remove the flower clusters carefully and discard the rest of the plant. Store the flower clusters in jars with tight-fitting tops, away from the sun.

Appendix: Dyeing with Natural Plant Materials

Plants have been the source of some of the most exquisite textile colors ever known. The magnificent costumes of the courts of Europe, the tapestries and church vestments, the jewel-like brilliance of the Oriental rugs from the Near and Far East, the subtlety and range of shades and tones in the textiles of the Chinese and Japanese—for thousands of years, all the dyes used to color fabrics and yarns were (with the exception of a very few shellfish and insects) obtained solely from plants.

Chemical dyes derived from coal tar were introduced during the middle of the nineteenth century and almost overnight made vegetable dyes obsolete. The production of these chemical colors has become a huge industry, and now almost any shade at all can be obtained merely by pouring it out of a bottle.

Recently, however, there has been a reawakening of interest in the ancient use of dyes made from plants, and craftspeople all over the world are joyfully rediscovering this craft. The process by which the various roots, barks, flowers, berries, and leaves yield their colors requires some effort, but the results are worth it.

Instructions and directions in this Appendix are meant to be a simple and satisfying introduction to this craft. Equipment and supplies have been kept at a minimum, and no special space is required.

All recipes are based on four ounces of natural, undyed wool yarn—wool because it is the easiest fiber to work with. Four-ounce amounts are readily obtainable, sufficient for adequately judging results, as well as a worthwhile quantity to save for a later use.

Mordants are used to "fix" the color of the dye in the yarn so that it will not run or fade. A number of chemicals and combinations of chemicals can be used. The recipes included here make use of the *alum/cream of tartar mordant.* This mordant is a particularly good one for the novice because it is simple to use and gives excellent, dependable results.

Equipment Needed

Soft water is preferred, and rain water is usually collected for this purpose. I myself have had good results with my own tap water. It comes from an artesian well and has chemical softeners added to it.

Four- or five-gallon enamel pot or stainless steel preserving kettle.

Rods for stirring—glass rods, wooden dowels, even chopsticks can be used.

Plastic bucket, helpful for rinsing yarn.

Plastic measuring spoons, for measuring mordant chemicals.

Notebook, to use for making notes on the work, is most helpful to have on hand.

Preparing the Yarn

Yarn must be washed before it is mordanted and dyed. To do this, untwist the skein but do not untie it, or it will tangle. Dissolve mild soap (never use detergent) in lukewarm water. Immerse the yarn, gently squeezing soap through. Rinse thoroughly in lukewarm water, squeezing out excess water gently. Never wring woolen yarn. Let dry.

Mordanting the Yarn

The mordant chemical potassium alum is available in drugstores, usually in four-ounce bottles. (It is used as an astringent for minor cuts, actually.) Cream of tartar is available in the baking section of markets.

For best results, yarn should be mordanted before it is dyed. To mordant four ounces of yarn, dissolve $1\frac{1}{2}$ tablespoons of potassium alum and $\frac{1}{2}$ tablespoon of cream of tartar in 1 cup of boiling water. Stir until fully dissolved, then add it to 2 gallons of tepid water in the kettle.

Saturate washed yarn in tepid water, squeeze out excess, and put it in the kettle. Heat slowly, and simmer gently for 1 hour. Do not let the water boil—this will cause the yarn to shrink. Remove from heat and let the yarn cool in the mordant bath. When the bath is cool, remove the yarn and gently squeeze out excess liquid. Place the damp yarn in a plastic bag and seal the bag. Mordanted yarn must be kept damp, and for this reason it is best to plan on dyeing the yarn immediately after mordanting, and certainly no more than a day or two after.

Premordanted yarn is added damp to the tepid dyebath. The dyebath is then slowly reheated, and kept at a simmer until the desired color or shade is obtained. The yarn should always be completely immersed in the dyebath. Stir it from time to time, and lift out the yarn occasionally to check how the color is developing. When the yarn is the shade you desire (bear in mind that colors look somewhat darker when they are wet), remove from the dyebath. Rinse in lukewarm water until the water runs clear. Let the dyed yarn dry away from the sun.

Dyeing with plant materials is not meant to be an exact science, but rather a fairly flexible craft. The plant materials are exciting to experiment with, and colors depend on a variety of factors. The actual amount of dye principle, as well as its strength and quality, depend to a great extent on the time of the year plants are collected, the weather during the growing season, and the condition of the plants themselves. In addition, the color depends on the amount of plant used, and the length of time the yarn is kept in the dyebath. Different mordants also influence colors. Occasionally the color or shade will not be precisely what you expected, but you will rarely, if ever, be disappointed—nature's colors are always special, and never quite like chemically dyed yarn from the store.

Should you wish to become further involved in this craft, several books on natural dyeing have been included in the Selected Bibliography, page 173.

Plants to Use for Natural Dyeing

Dyes can be obtained from the following plants in this book:
BLACKBERRY: Terra cotta from the leaves, purple-brown from the ripe berries.

DANDELION: Light yellow from the flowers.

DOCK: Yellow-gold to tan from the roots.

GOLDENROD: The flowers yield various yellows and yellow-greens.

LADY'S-THUMB: Green from the whole plant.

POKEWEED: Deep rose to light pink from the ripe berries, but colors are not fast in sunlight.

QUEEN ANNE'S LACE: Pale yellow to pale yellow-green from the whole plant.

WILD SUMAC: Tan to brown from the leaves and young shoots. No mordant required.

Preparation of Plant Material

Unless otherwise noted, dye principles are extracted from the plants as follows:
Leaves and young shoots are chopped and soaked in water to cover for 3 hours.
Berries should be gently crushed and soaked in water to cover for 2 hours.
Roots should be washed well, finely chopped, and soaked in water overnight.
Whole plants should be chopped and soaked in water to cover for 3 hours.

Preparation of the Dyebath

After the plant material has been soaked, boil it in the soaking water for 1 hour to fully extract the dye, then let the dyebath cool to lukewarm. Strain the dyebath before adding the yarn if you have used berries. It is not necessary to remove leaves, flowers, or roots. Any bits that cling to the yarn wash away when it is rinsed.

Premordanted, damp yarn is added to the lukewarm dyebath, never when it is hot. The yarn is gradually reheated in the dyebath until the boiling point is reached, and the time is counted from that point.

BLACKBERRY: To dye wool terra cotta, gather 1 gallon of leaves. After soaking, boil gently for 1 hour. Add yarn and simmer at least 30 minutes or until desired color is obtained.

To dye wool purple brown, gather about 3 quarts of ripe berries. After soaking, simmer between 45 minutes and 1 hour. Strain out berries. Add yarn and simmer for about 40 minutes, or until desired color is obtained.

DANDELION: To dye wool pale yellow, gather about 3 quarts of freshly opened flowers. After soaking, simmer gently for 45 minutes. Add yarn, and simmer approximately 40 minutes or until desired color is obtained.

DOCK: To dye yarn dark yellow to gold, gather enough roots to yield 2 cups when chopped. For darker tones, use more root material and leave yarn in the dyebath longer. After soaking, boil about 35 minutes, add yarn, and simmer at least 20 minutes or until desired color is obtained.

GOLDENROD: The flowers of all the Goldenrods yield an excellent dye of varying shades of yellow. They are best gathered just as they start to open, for at this time the color-producing principle is strongest. To dye wool golden yellow to yellow-green, gather about 2 quarts of freshly opened flowers. Cover the flowers with water and simmer for 45 minutes. Add yarn and simmer 40 minutes or until desired color is obtained.

LADY'S-THUMB: To dye wool green, gather 2 quarts of the whole plant, except roots. After soaking, boil gently 45 minutes. Add yarn and simmer for 1 hour, or until desired color is obtained.

POKEWEED: The beautiful rose-pink colors obtained from ripe Pokeberries are unfortunately not fast, but you might want to do some experimenting with this plant. To dye 4 ounces of yarn, gather 2 quarts of ripe berries. After soaking, boil gently for 30 minutes. Strain, add yarn, and simmer until desired color is obtained. The range is from deep rose to pale pink.

QUEEN ANNE'S LACE: To dye wool pale yellow to yellow-green, gather 4 quarts of the whole plant (except the roots). After soaking, boil gently for 45 minutes. Add yarn and simmer about 45 minutes to 1 hour, or until desired color is obtained.

WILD SUMAC: Sumac contains a substantial amount of tannin, a natural mordant. Therefore, premordanting yarn is not necessary in this instance. To dye yarn tan to dark brown, gather about 1 quart of fresh leaves—more for a darker color. Soak at least 12 hours. After soaking, bring to a boil and simmer for 45 minutes. Add washed yarn and simmer gently until desired color is obtained.

Glossary

ALTERNATIVE: An agent tending to gradually alter a condition.

CARMINATIVE: An agent for warming the stomach and relieving gas in the stomach and intestinal tract.

CATHARTIC: An agent capable of causing evacuation from the bowels. There are two kinds of cathartics: *aperients,* or *laxatives,* are mild acting. *Purgatives* are very strong.

DECOCTION: Decoctions are prepared by boiling botanical materials, usually (but not always) seeds, roots, and barks, in water to extract their medicinal properties. The resulting liquid is a decoction.

DEMULCENT: Any agent of a bland, usually mucilaginous nature, used *internally* for its soothing properties. Some demulcents are capable of forming a protective coating.

DIAPHORETIC: An agent that is capable of increasing perspiration.

DIURETIC: An agent that increases the flow of urine.

EMOLLIENT: An agent of a mucilaginous or oily nature, used *externally* for its soothing and softening properties.

EXPECTORANT: An agent used to cause expulsion of phlegm from the bronchial and nasal passages.

INFUSIONS: Infusions are prepared by pouring boiling water over plant materials, usually leaves or flowers, and allowing them to steep. The resulting liquid is an infusion.

NARCOTIC: An agent that relieves pain and induces sleep. Large doses may cause coma; overdoses can result in death.

REFRIGERANT: A cooling agent, usually a beverage.

SEDATIVE: An agent used to ease irritation, discomfort, and pain.

STIMULANT: An agent capable of increasing or quickening a variety of functional actions of the body.

TONIC: Any agent that stimulates the appetite and aids digestion.

VULNERARY: An agent that promotes the healing of minor wounds.

Selected Bibliography

Allport, Noel. *The Chemistry and Pharmacy of Vegetable Drugs*. London: George Newnes, Ltd., 1943.

Angier, Bradford. *Free for the Eating*. Harrisburg, Pa.: Stackpole Books, 1966.

Arena, J., and J. Garden. *Human Poisoning from Native and Cultivated Plants*. Durham, N.C.: Duke University Press, 1969.

Bailey, L. H. *Standard Cyclopedia of Horticulture*. New York: The Macmillan Co., 1933.

Baker, M. *Discovering the Folklore of Plants*. London: Shire Publications, 1971.

Bergen, Fanny D. *Animal and Plant Lore from the Oral Tradition of English-Speaking Folk*. Boston: Houghton Mifflin Co., 1899.

Blackwell, Elizabeth. *A Curious Herbal*. London: printed for J. Nourse, 1737.

Brown, Annora. *Old Man's Garden*. Toronto: J. M. Dent & Sons, Ltd., 1954.

Cocannouer, Joseph. *Weeds, Guardians of the Soil*. New York: Devin-Adair, 1950.

Cockayne, Oliver. *Leechdoms, Starcraft and Wortcunning of Early England*. Chronicals and Memorials of Great Britain and Ireland During the Middle Ages, Number 35. London: Great Britain Public Records Office, 1864.

Coles, William. *Adam in Eden*. London: Nathaniel Brooke, 1657.

———. *The Art of Simpling*. London: Nathaniel Brooke, 1656.

Coon, Nelson. *Using Plants for Healing*. New York: Hearthside Press, 1963.

Crow, W. B. *The Occult Properties of Herbs*. London: Aquarian Press, 1969.

Crowhurst, Adrienne. *The Weed Cookbook*. New York: Lancer Books, 1972.

Dowden, Anne Ophelia. *Wild Green Things in the City*. New York: Thomas Y. Crowell Co., 1972.

Evelyn, John. *Acetaria: A Discourse on Sallets*. London, 1699.

Fernald, M. L., and A. C. Kinsey. *Edible Wild Plants of Eastern North America*. New York: Harper & Row, 1958.

Folkhard, Richard. *Plant Lore and Legends*. London: Sampson, Low, Marston, Searle & Rivington, 1884.

Gerard, John. *Herball*. London, 1633.

Gibbons, Euell. *Stalking the Healthful Herbs.* New York: David McKay, 1966.

———. *Stalking the Wild Asparagus.* New York: David McKay, 1970.

Godshall, Ammon. *Edible, Poisonous and Medicinal Fruits of Central America.* Panama Canal, 1942.

Green, Thomas. *The Universal Herbal.* London, 1823.

Grieve, Maude. *A Modern Herbal.* New York: Hafner Publishing Co., 1957.

Grimm, William C. *Recognizing Native Shrubs.* Harrisburg, Pa.: Stackpole Books, 1966.

Hamarneh, Sami. *Origins of Pharmacy and Therapy in the Near East.* Tokyo: Niato Foundation, 1973.

Harris, Ben Charles. *Eat the Weeds.* Barre, Mass.: Barre Publications, 1968.

Hatfield, Audrey Wynne. *How to Enjoy Your Weeds.* New York: Sterling Publishing Co., 1971.

Heiser, Charles B., Jr. *Nightshades, the Paradoxical Plants.* San Francisco: W. H. Freeman & Co., 1969.

Kingsbury, John M. *Poisonous Plants of the United States and Canada.* Englewood Cliffs, N.J.: Prentice-Hall, 1964.

———. *Deadly Harvest: A Guide to Common Poisonous Plants.* New York: Holt, Rinehart and Winston, 1965.

Klimas, John E., and James A. Cunningham. *Wildflowers of Eastern America.* New York: Alfred A. Knopf, 1974.

Krochmal, A., and P. W. LeQuesne. *Pokeweed.* USDA Forest Service Research Paper, NE 177. Upper Darby: Northwestern Forest Extension Service, 1974.

Langham, William. *The Garden of Health.* London, 1633.

Lehner, E., and J. Lehner. *Folklore and Odysseys of Food and Medicinal Plants.* New York: Tudor Publishing Co., 1962.

Lovelock, Yann. *The Vegetable Book: An Unnatural History.* New York: St. Martin's Press, 1968.

Martin, Alexander C. *Weeds.* New York: Golden Press, 1972.

Medsger, Oliver P. *Edible Wild Plants.* New York: Collier Books, 1939.

Meunscher, Walter C. *Poisonous Plants of the United States.* New York: The Macmillan Co., 1970.

———, and Myron A. Rice. *Garden Spice and Wild Potherbs.* Ithaca, N.Y.: Cornell University Press, 1955.

Meyer, Joseph D. *The Herbalist.* Chicago: Rand McNally & Co., 1960.

Millspaugh, Charles F. *American Medicinal Plants.* New York: Dover Publications, 1970.

Monroe, J. *The American Botanist and Family Physician.* Vermont: Jonathan Morrison, 1824.

Morton, Julia. *Folk Remedies of the Low Country.* Miami, Fla.: E. A. Seeman, 1974.

Oldmeadow, K. *The Folklore of Herbs.* Birmingham, Eng.: Cornish Brothers, 1946.

Orr, Robert T., and Margaret Orr. *Wildflowers of Western America.* New York: Alfred A. Knopf, 1974.

Page, Nancy M., and Richard E. Weaver, Jr. *Wild Plants in the City.* New York: Quadrangle Books/The New York Times Book Co., 1975.

Parkinson, John. *Paradise in Sole.* London, 1635.

———. *Theatre of Plants.* London, 1640.

Peterson, Roger T., and M. McKenny. *Field Guide to the Wildflowers of Northeast America.* Boston: Houghton Mifflin Co., 1968.

Pfeiffer, Ehrenfried. *Weeds and What They Tell Us.* Emmaus, Pa.: Rodale Press, 1945.

Porter-Smith, E., and G. Stuart. *Chinese Medicinal Herbs.* San Francisco: Georgetown Press, 1973.

Potter, S. *Therapeutics, Materia Medica and Pharmacy.* 13th ed. Philadelphia: P. Blakiston's Son & Co., 1917.

Princess Atalie. *The Earth Speaks.* New York: Fleming H. Revell Co., 1940.

Quinn, Vernon. *Vegetables in the Garden and Their Legends.* Philadelphia: J. P. Lippincott Co., 1942.

Radford, E., and M. A. Radford. *Encyclopedia of Superstitions.* London: Huchinson, 1961.

Rafinesque, C. S. *Manual of Medical Botany of North America.* Philadelphia: Atkinson and Alexander, 1828.

Savage, F. G. *Flora and Folklore of Shakespeare.* London: E. J. Burrow and Co., Ltd., 1923.

Skinner, Charles. *Myths and Legends of Flowers, Trees, Fruits and Plants.* Philadelphia: J. P. Lippincott & Co., 1925.

Swain, Tony, ed. *Plants in the Development of Modern Medicine.* Cambridge, Mass.: Harvard University Press, 1972.

Tantaquidgeon, Gladys. *Folk Medicine of the Delaware and Related Algonkian Tribes.* Harrisburg, Pa., Pennsylvania Historical Museum Commission, 1972.

Turner, William. *A New Herbal.* London: Arnold Birckman, 1551.

Weiner, Michael A. *Earth Medicine—Earth Foods: Plant Remedies, Drugs and Natural Foods of the North American Indians.* New York: Collier-Macmillan, Ltd., 1972.

Books on Dyeing with Natural Plant Materials

Duncan, Molly. *Spin Your Own Wool, Dye It and Weave It.* Sydney, Australia: A. H. & A. W. Reed, 1969.

Dye Plants and Dyeing. Vol. 20, No. 3. Brooklyn, N.Y.: The Brooklyn Botanic Gardens.

Grae, Ida. *Nature's Colors.* New York: The Macmillan Co., 1974.

Natural Plant Dyeing. New Series, Vol. 29, No. 2. Brooklyn, N.Y.: The Brooklyn Botanic Gardens.

Svinicki, Eunice. *Step by Step Spinning and Dyeing.* New York: Golden Press, 1974.

Index

A NOTE ABOUT THE AUTHOR

Maida Silverman was born and has spent her life in New York City. The fantasy of a garden of her own, overrun with fruit trees, nut trees, flowers, berries, vegetables, and herbs has not yet come to pass. Because of this she turned her attention to wild plants that grow in the city and found them rewarding and engrossing.

She graduated from Pratt Institute in 1962. In addition to being an illustrator, she is a craftswoman whose major field is stitchery. She often spins her own yarns and dyes them, using vegetable dyes derived from some of the plants described in A City Herbal. *Her illustrations include those for* The Poetry of Chaim Nachman Bialik *(published by the Union of American Hebrew Congregations), which was voted one of the Fifty Best Books of the Year by the American Institute of Graphic Arts in 1973. Her stitchery projects and writings on the use of herbs and wild plants have appeared in* Family Circle *and* Seventeen *magazines.*

A NOTE ABOUT THE TYPE

The text of this book was set in Intertype Garamond, a modern rendering of the type first cut by Claude Garamond (1510–1561). Garamond was a pupil of Geoffroy Tory and is believed to have based his letters on the Venetian models, although he introduced a number of important differences, and it is to him we owe the letter which we know as old-style. He gave to his letters a certain elegance and a feeling of movement that won for their creator an immediate reputation and the patronage of Francis I of France.

The book was composed by American Book–Stratford Press, Brattleboro, Vermont. Printed and bound by The Book Press, Brattleboro, Vermont. Designed by Gwen Townsend and Maida Silverman.